RESEARCH PROGRESS IN BIOTECHNOLOGY

Research Progress in Biotechnology

G. E. Zaikov
Editor

Nova Biomedical Books
New York

For permission to use material from this book please contact us:
Telephone 631-231-7269; Fax 631-231-8175
Web Site: http://www.novapublishers.com

Library of Congress Cataloging-in-Publication Data

Research progress in biotechnology / G.E. Zaikov (editor).
p. ; cm.
Includes bibliographical references and index.
ISBN 978-1-60456-000-8 (hardcover)
1. Biotechnology. I. Zaikov, Gennadii Efremovich.
[DNLM: 1. Biotechnology--methods. 2. Biomedical Engineering--methods. W 82 R432 2007]
TP248.2.R46 2007
660.6--dc22

2007035263

Published by Nova Science Publishers, Inc. ✢ New York

Contents

Preface

Biotechnology is a collection of technologies that capitalize on the attributes of cells and biological molecules. Biotechnology will help improve the ability to customize therapies based on individual genomics; prevent, diagnose, and treat all types of diseases rather than rely on rescue therapy and provide breakthroughs in agricultural production and food safety. This book presents the latest research in the field.

In: Research Progress in Biotechnology
Editor: G. E. Zaikov, pp. 1-5
ISBN: 978-1-60456-000-8

Chapter I

Biochemical Cycle of Synthesis of Arginine and Nitric Oxide in the Liver

L.M. Baider and Z.V. Kuropteva*
Emanuel Institute of Biochemical Physics of the RAS,
119991 Moscow, Kosygin str. 4, Russia

Studies performed in the two past decades showed that the simple molecule of nitric oxide (NO) is a versatile regulator of numerous biochemical processes in humans and animals. It is known that NO is synthesized from L-arginine and oxygen by special enzymes—NO synthases. Today, the most well-known are three isoforms of this enzyme: two constitutive (endothelial and neuronal NO synthases; eNOS and nNOS, respectively) and one inducible iNOS). The first two NO synthases, eNOS and nNOS, are usually activated by increased concentrations of cytosolic Ca^{2+}, which stimulates NO production within several minutes. These enzymes synthesize small quantities of NO (no more than tens micromoles). The inducible isoform of NO synthase, which is expressed predominantly in immunocompetent cells and nearly in all cells of the liver after stimulation with mediators of immune or inflammatory reactions, produces NO at millimolar concentrations [1]. In these cells, NO is used as an effector molecule that is involved in elimination of various pathogens (bacteria, viruses, tumor cells, etc.). A characteristic feature of iNOS in these cells is that its induction (and, therefore, NO synthesis) continues for a long time after stimulation—for several hours or even days.

For example, in experimental models of endotoxemia *in vivo,* iNOS mRNA expression was detected for 3 h after the addition of a stimulator, whereas the maximal accumulation of iNOS was observed after 12-16 h [2]. In human hepatocytes stimulated by cytokines and LPSs, a single band of mRNA was detected as early as3 h after stimulation; it reached maximum 8 h after stimulation and became negligible 48 h after stimulation. The level of mRNA was compared to the activity of iNOS, which was estimated by the accumulation of NO_2^- and NO_3^- in culture supernatant. The amount of these anions increased in 20-30 times 24 and 48 h after stimulation [3]. It was shown that the induction of iNOS in macrophages, neutrophils, and

* Z.V. Kuropteva: e-mail: zvk@sky.chph.ras.ru

E. coli may be caused by ascorbic acid and that one peritoneal macrophage can produce as many as 1.5 x 10^9 NO molecules [4-6]. With regard for the fact that one arginine molecule is consumed for the synthesis of one NO molecule, a question arises as to where such a substantial amount of arginine can be taken. At such a high level of consumption, arginine should be synthesized as required. In mammals, arginine is synthesized in the liver in the urea cycle from citrulline and ammonia. It is believed that thus formed arginine is rapidly cleaved by arginase to urea and ornithine [7]. Published data on the withdrawal of arginine from the urea cycle and its function as a substrate of iNOS are absent [1,8,9]. In view of this, a question arises as to how the organism satisfies its increased requirements for arginine. In this study, we obtained data suggesting the existence of a special cycle of synthesis of arginine and NO in liver cells.

Materials and Methods

Ammonium chloride, citrulline, and arginine used in the study were from "Acros" (United States); interferon-gamma from the company "Ferment" (Russia). Experiments were performed with isolated liver of SHK mice. After decapitation of animals, the liver was extracted, ground to small pieces, and added to saline. This mixture was divided into three portions. One portion was supplemented first with interferon-gamma at a dose of 100IU; after a 30-min incubation, ammonium chloride and citrulline (2 x 10^{-3} M each) were added. The second portion was incubated only with interferon-gamma. The third portion served as a control; it contained no additives and was incubated at room temperature under the same conditions. After 3 and 19 h of incubation, the pieces of liver were placed into teflon tubes, frozen in liquid nitrogen, and then pushed out of the tubes. Thus obtained specimens representing frozen columns (diameter, 3 mm; height, 30 mm) were analyzed by EPR.

EPR spectra of these columns were recorded in an ESP-300 Bruker Analitische Messtechnik spectrophotometer (Germany) at a temperature of liquid nitrogen.

Results and Discussion

As mentioned above, the functioning of the inducible NO synthase, iNOS, results in a long-term production of high quantities of NO; this process is highly arginine-consuming, because one molecule of arginine yields one molecule of NO. It is believed that the arginine synthesized in the urea cycle does not escape from the cycle. Nothing is known on the stores of arginine in the organism. Therefore, it should be synthesized *de novo* in accordance with increasing demands for it.

We studied the effect of the urea cycle participants—ammonium chloride (a well-known activator of the urea cycle) and citrulline—on the synthesis of NO in liver cells. Liver tissues are a convenient object of such a study, because almost all liver cells (hepatocytes, as well as Kupffer, Ito, and endothelial cells) are capable of iNOS induction and NO synthesis [1]. Liver cells were first stimulated with interferon-gamma and then treated with ammonium chloride and citrulline. NO production was monitored by the appearance of nitrosyl complexes heme-NO in EPR spectra of liver tissue samples. Additional NO traps were not introduced, because the natural traps of blood and liver cells—heme proteins—were used. It is known that heme proteins (hemoglobin and cytochrome P-450 in the liver) have a high affinity for NO and, upon accepting it, form stable nitrosyl complexes heme-NO, which determine a broad EPR spectrum with $g = 2.02$ and a triplet structure with a splitting 17 G at $g = 2.01$. Figure 1 shows EPR spectra of 1 samples incubated at room temperature in the presence of interferon-gamma, ammonium chloride, and citrulline (spectrum 1) and in the presence of interferon-gamma alone (spectrum 2). In normalcy, EPR spectra of liver specimens contained the signals from the cytochrome P-450 ($g_1 = 2.42$ and $g_2 = 2.25$) and the signals determined by components of the mitochondrial respiratory chain: the singlet signal from flavosemiquinones in the free radical range and the signal from the Fe-S centers of the NADH-dehydrogenase complex (centers *N*-1b) with $g = 1.94$. The EPR spectra of liver specimens incubated in the presence of interferon-gamma, ammonium cloride, and citrulline (figure 1) contained the components of the known signal with the triplet splitting at g 2.01 from the complexes of heme proteins with NO (heme-NO). It was made possible to resolve this signal by subtracting the EPR spectra of the liver specimens incubated solely in the presence of interferon-gamma (figure 2) from the EPR spectra of the liver specimens incubated in the presence of additives under the same conditions. The appearance of the heme-NO complexes in the spectra (figure 1) indicates that NO was synthesized in the presence of ammonium chloride and citrulline in the liver cells that were preliminarily activated with interferon-gamma. Note that no increase in NO production was observed in the first hours after the addition of ammonium chloride, an activator of the urea cycle, both simultaneously with citrulline and in the absence of the latter. This finding indicates that the activation of the urea cycle does not lead to additional NO production in the firth three hours after activation of the cycle and that NO is not synthesized without iNOS induction. On the basis of these data, it was logical to assume that, since citrulline is produced along with NO during iNOS functioning, it can be used again for the synthesis of arginine. At a sufficient amount of the source of nitrogen (in our case, ammonia), the same enzymes that function in the urea cycle (argininosuccinate synthase and arginine synthase) may synthesize arginine. Hence, in addition to the urea cycle, there exists a cycle of synthesis of arginine in liver cells, which is shorter and more labile than the urea cycle. This cycle is apparently coupled to the urea cycle; it is triggered only in the case of iNOS induction and can be activated by the substrates of the enzyme that synthesizes arginine (figure 2). This cycle can be termed the minor arginine cycle, or the cycle of synthesis of arginine and nitric oxide. It consists of iNOS, synthesizing NO from arginine, and the enzymes that synthesize arginine from citrulline and the source of nitrogen (e.g., ammonia or glutamine). This cycle functions when iNOS is expressed and cells require increased amounts of NO. It is likely that the use of arginine for the synthesis of NO has certain advantages over its use in the urea cycle, because NO is a key effector molecule in the system of cell immunity. This minor cycle is a shunt for the urea cycle.

We studied the possibility of existence of an independent cycle of synthesis of arginine and NO in peritoneal macrophages as well. The date were obtained showing that a cycle of synthesis of arginine and NO apparently functions in macrophages, similarly to liver cells [10].

Thus, the results of this study suggest that an independent cycle of synthesis of arginine and nitric oxide may function in the cells expressing inducible NO-synthase. The cycle allows the cells to synthesize NO for a long time in the quantities required. Therefore, the process NO synthesis from L-arginine apparently is not linear and represents a closed cycle, in which citrulline produced in the course of NO synthesis (which regarded earlier a by-product) is used again for the synthesis of arginine and then NO. The amount of produced NO will be limited by the presence of substrates of the enzymes that catalyze the synthesis of arginine (in addition to citrulline) and coenzymes of iNOS, The latter, along with other NO synthases, is a precisely regulated enzyme whose normal functioning requires the presence of six cofactors: FMN, FAD, NADH, BH_4, NADPH, and calmodulin bound in its active center. Earlier, we discovered that ascorbic acid is able to stimulate NO production in leukocytes and macrophag es [4-5]. It is most likely that this phenomenon is also due to the functioning of this cycle. Ascorbic acid may have been involved in the maintenance of iNOS cofactors (in particular, NADPH and BH_4) in working reduced state.

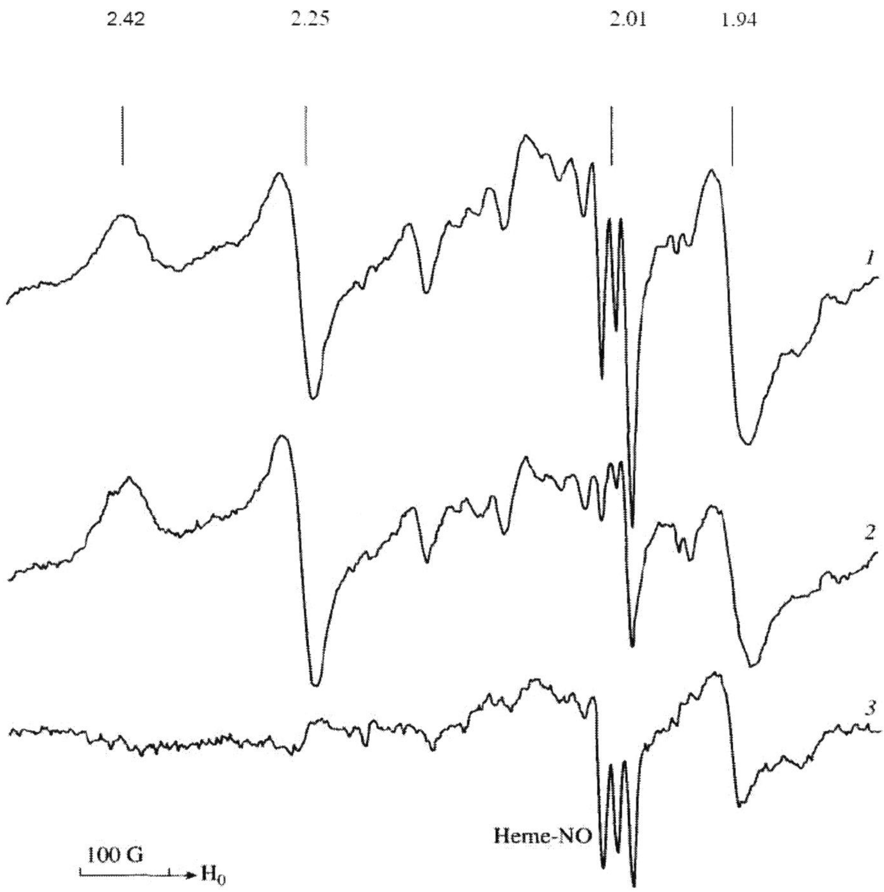

Figure 1. EPR spectra of liver specimens incubated at room temperature in the presence of (1) - interferon-gamma, ammonium chloride, and citralline and (2) – interferon-gamma alone. Figure 3 shows the differential spectrum obtained by subtracting spectrum 2 from spectrum 1. Spectra were recorded under the following conditions: microwave power, 20 mVt; amplitude of magnetic field modulation ,4 G; temperature, 77 K.

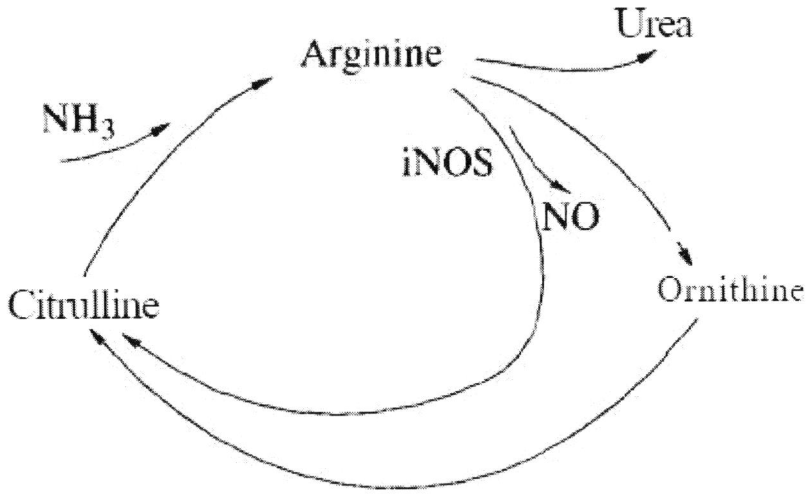

Figure 2. Simplified scheme of intertwining urea cycle and the putative cycle of NO synthesis, in which inducible NO synthase is involved.

References

[1] Teilor, B.S., Alarson, L.Kh., and Billiar, T.R., *Biokhimiya,* 1998, vol. 63, issue 7, pp. 901-922.

[2] Geller, D.A., Freeswick, P.D., Nguyen, D., *et al., Arch. Surg.,* 1994, vol. 129, pp. 165-171.

[3] Nussler, F., Disilvio, M., Billiar, N.R., *et al, J. Exp. Med,* 1992, vol. 176, pp. 261-264.

[4] Kuropteva, Z.V., Baider, L.M., and Zhumabaeva, T.T., *Biofizika,* 2000, vol. 45, issue 4, pp. 671-674.

[5] Kuropteva, Z.V., Zhumabaeva, T.T., Baider, L.M., and Aleshchenko, A.V., *Dokl. Akad. Nauk,* 2001, vol. 376, no. 2, pp. 258-261 *[Dokl. Biochem. Biophys.* (Engl. Transl.), vol. 376, no. 2, pp. 7-9].

[6] Zhumabaeva, T.T., Baider, L.M., Volodina, L.A., and Kuropteva, Z.V., *Prikl. Biokhim. Mikrobiol,* 2001, vol. 37, no. 6, pp. 742-746.

[7] Leninger, A.L., *Principles of Biochemistry,* New York: Worth, 1982. Translated under the title *Osnovy biokhimii,* Moscow: Mir, 1985, vol. 2, pp. 589-594.

[8] Wettstein, M., Gerok, W., and Haussinger, D., *Hepatology,* 1994, vol. 19, pp. 641-647.

[9] Pastor, CM., Morris, S.M., and Billiar, T.R, *Am. J. Physiol,* 1995, vol. 269, pp. G861-G866.

[10] Baider L.M., Aleshchenko A.V., Kuropteva Z.V. ., *Dokl Akad. Nauk,* 2005, vol. 405, no. 4, pp. 547-549 *[Dokl Biochem. Biophys.* (Engl. Transl.), vol. 405, pp. 429-431].

In: Research Progress in Biotechnology
Editor: G. E. Zaikov, pp. 7-14
ISBN: 978-1-60456-000-8

Chapter II

The Biocomplex which Decreases Damage of Normal Tissues During Chemotherapy or Radiation Therapy but without Protective Action on Tumor

T.G. Malinina[*1], *E.Yu. Lyubimova*[1],
O.A. Konoplyannikova[*2], *and A.G. Konoplyannikov*[2]
[1] The Medical Scientific and Production Complex "Biotechindustry",
Moscow, B. Communisticheskaya, 11-2, 117447, Russia
[2] Department of Cell Therapy and Experimental Radiation Therapy of the State Institution of Medical Radiological Research Center of RAMS,
4 Korolev Str., Obninsk, Kaluga Region, 249036, Russia

Abstract

Medical Scientific and Production Complex "Biotechindustry" continues the studies of relation of structure, properties and functions in placentary complexes (PC) for the further decision of modern medicine practical problems. Earlier we have found out that PC may to activate healing of a wound and to render radioprotective, antistressful and immunomodulated action. In this report we inform about the biologically active food additive "Gastrofilin" (combination of biologically active substances from PC and some plants) that is capable to decrease damage of normal tissues during chemotherapy or radiation therapy but without protective action on tumor growth and formation of metastasizes. Thus, we have developed a new opportunity to use of placentary complexes for protection of "critical" normal tissues radical therapy of tumor resistant forms.

* e-mail:biid@km.ru; Tel./fax☹495) 911 90 79
* e-mail: konopl@mrrc.obninsk.ru, Tel.: +4 48439 93290; fax: +4 495 9561440

Keywords: radiation therapy, chemotherapy, biocomplex "Gastrofilin"(GF), hematopoietic steam cells, Lewis lung carcinoma cells, skin reactions.

Introduction and Objective

Last twenty years attention of some scientists and medical experts began to attract more and more biological active substances of animals and plants derivation. "Lutex" is a natural multi-component, multi-functional system of biologically active substances, roduced on a basis of the new biotechnology for processing of placenta, developed by Russian research center of "Medical Scientific and Production Complex Biotechindustry"(1). This multicomponent preparation contains: amino acids (including essential amino acids), low molecular weight peptides, proteins, nucleic acids, cytokines (interleukins 1β and 6, γ– and β–interferons, growth factors – epidermal growth factor (EGF), fibroblast growth factor (FGF), hialuronic acid (2), lipids, Q10, alpha-fetoprotein, polyunsaturated fatty acids, 2-dimethylaminoethanol (DMAE), organic acids, glutathione, balanced complex of vitamins, macro- and microelements. It is also powerful antioxidant (3). «Lutex» is the main part of biologically active food additive «Gastrofyline» (GF) with applied in complex therapy of some diseases of upper gastrointestinal tract and liver. It is effective and safe agents protecting from the action of various extreme factors. The radioprotective properties of the placental complexes were studied in experimental conditions (1, 3). This complex protected mice during oral administration before irradiation. In optimal dose of GF applied 30 min before whole body lethal dose of γ-irradiation in 7.5 Gy increased the survival of mice up to 75 %. It was discovered that placental complexes stimulated proliferation activity of hematopoietic stem cells in vivo and its circulation in blood (4). Thus, GF can be perspective biocomplexes for application in modern practical medicine.

Now in the modern medicine such radioprotective preparations are only two or three, but all of them have restrictions for application, as are accompanied by side-effects (vomiting, temporary loss of consciousness).

The further researches have been devoted to studying of an opportunity of GF application in complex therapy of cancer diseases (radiation therapy and chemotherapy).

Our studies were spent in several directions:

- Protective effects of GF on mouse hematopoietic stem cells (CFU-S) during irradiation or action of cyclophosphamide in vivo or for mouse skin reactions after irradiation;
- Effects of GF on sensitivity of mouse transplanted tumors cells – Lewis lung carcinoma to local irradiation or action of cyclophosphamide in vivo.

Materials and Methods

Radioprotective and hemostimulating action of GF we studied using of methods for testing of hematopoietic stem cells - CFU-S-8, that forming of endogenous and exogenous spleen colonies at mice-hybrids F1(CBAxC57Bl/6), male, 3 months old (5, 6, 7). At use of the endogenous spleen colonies method we studied effect of different doses GF administrated *per os* on mice one day before total γ-irradiation (Co^{60}) in dose 6 Gy. There were 15 mice in each experimental group. On 8th day after γ-irradiation the spleens of mice in different groups were weighed and fixed in a mix of ethanol and ice acetic acid (1:4) during 2-3 hours. After fixing of spleens we counted up number of colonies on spleen surface which diameter was equal or there were more than 0.4 mm. The mean number of endogenous spleen colonies and a standard error were calculated on the basis of representation that this set is described by log-normal distribution [7, 8]. At use of the exogenous spleen colonies method we studied effect of GF (25 ml/kg) administrated *per os* on mice-donors of bone marrow one day before preparation of femoral bone marrow or before injection of cyclophosphamide (200 mg/kg) and the subsequent preparation of bone marrow (2 hours after injection of cyclophosphamide). The bone marrow at these mice-donors was received from mouse femur and bone marrow cells suspension was prepared on PBS with concentration $5x10^5$ cells/ml. Preliminary irradiated in lethal dose 8.5 Gy mice-recipients has been injected intravenously 0,2 ml prepared bone marrow cells suspension for each experimental group. On 8th day after γ-irradiation of mice-recipients and transplantation of bone marrow cells the spleens of these mice were weighed and fixed as for methods of endogenous spleen colonies. The mean number of exogenous spleen colonies and a standard error were calculated on the basis of representation that this set is described by normal distribution (8). In this series of experiences on each experimental group we used 5 mice (F1(CBAxC57Bl/6), male, 3 months old) as donor of bone marrow cells, and 15 the some strain mice-recipients of bone marrow cells.

The technique of an estimation of the skin damages caused by local γ-irradiation of hind leg of mice (F1(CBAxC57Bl/6), male, 3 months old) is based on the mark rating scale of various skin reactions (hardness, dry peeling, exudative epidermitis), offered in work (9) and modified in our laboratory (10). Supervision of skin reaction in this case was spent within 4 weeks after local γ-irradiation (Co^{60}) in dose 12 Gy, and the quantitative estimation of skin reaction in conventional units was given on the basis of calculation of the area under the curve of increase and easing of radiation reaction (10). There were 12 mice in each experimental and control groups.

In experiences for studying GF action on sensitivity of tumor cells to ionizing radiation or cyclophosphamide we used resistant form of transplanted in vivo tumor – Lewis lung carcinoma (LLC). 10^6 cells of LLC were transplanted to mice, stain C57Bl/6, male, 3 months old, under skin of femur. On day 8th after tumor cells transplantation into mice we locally irradiated tumors (with mean volume in 45 mm^3) in dose 10 Gy (Co^{60}) or subjected to 4 day time rate of therapy by cyclophosphamide (the total dose of drug was 200 mg/kg). On day 22th after tumor cell transplantation we measured tumor volume on mouse leg and estimated number of tumor metastasises in lungs (on lung surface after tissues fixation by a mix of ethanol and an ice acetic acid as for spleen colonies). In these experiences in each control and experimental groups was on 15 mice.

The validity of difference between experimental and corresponding control groups we estimated using Student t-criterion.

Results and Discussion

Radioprotective action of GF applied in various concentrations and administrated *per os* into mice-hybrids one day before γ-irradiation in dose 6 Gy is presented in table 1. Expressed radioprotective effect of GF on the level of formed by CFU-S endogenous colonies in mice spleens is observed after application of this preparation in dose of 25 ml/kg and more. By method of spleen exogenous colonies it has been shown, that application of GF in a dose 25 ml/kg one1 day before testing increased approximately in one and a half time the contents of CFU-S in population of donor bone marrow cells (table 2). The reasons of this effect remains not enough investigated, it is possible to suggest, that GF promotes start in proliferation of resting hematopoietic stem cells just as it is observed for many biological reactions modifiers [11]. Alongside with radioprotective action GF it is capable to decrease damage of CFU-S caused by injection into mice anticancer drug - cyclophosphamide (table 3).

In table 4 it was presented data about radioprotective action of GF on skin of hind leg of mice irradiated locally in dose 12 Gy. This reaction was estimated by modified method of study for acute radiation reaction of animal skin (9, 10). Statistically significant radioprotective effect of GF application on mouse skin has been revealed for dose this agent measuring 40 ml/kg.

In tables 5 and 6 there were presented data about effects of GF application on effectiveness of radiation therapy and chemotherapy on growth and metastasizing of Lewis lung carcinoma cells in vivo. Statistically significant protective action of GF on tumor growth and formation of lung metastasizes in this study is lacking.

Table 1. Action of Gastrofyline administrated *per os* on mice-hybrids F1(CBAxC57Bl/6), male, 3 months old, in various concentration one day before total γ-irradiation in dose 6 Gy on number of endogenous spleen colonies, formed by CFU-S on day 8th after irradiation. There were 15 animals in each group

№	Groups	Mean weight of spleen (M±m), mg	Mean number of endogenous spleen colonies, (M±m)
1	Gastrofyline, 40 ml/kg + 6 Gy	54.4 ± 3.2 *	5.5 ± 0.6 *
2	Gastrofyline, 25 ml/kg + 6 Gy	50.2 ± 2.9 *	5.1 ± 0.6 *
3	Gastrofyline, 5 ml/kg + 6 Gy	46.8 ± 3.0	3.2 ± 0.4 *
4	Control, 6 Gy	43.1 ± 2.1	1.9 ± 0.4

* - a significant difference with the control group ($p < 0.05$).

Table 2. Influence of Gastrofyline (25 ml/kg) administrated *per os* one day before preparation of bone marrow cells (donor-mice) or gamma-irradiation in dose 8.5 Gy (recipient-mice) into the same mice (mice-hybrids F1(CBAxC57Bl/6), male, 3 months old) on level of exogenous spleen colonies, formed by CFU-S on day 8th after i/v transplantation of bone marrow cells (10^5 cells) in recipients. There were 5 mice in each donor group and 15 mice in each recipient group

№	Groups	Mean weight of spleen (M±m), mg	Mean number of exogenous spleen colonies, (M±m)
1.	Control of recipient mice irradiated in dose 8.5 Gy	36.4 ± 1.9	>0.2 ± 0.1
2.	Administration of Gastrofyline in recipient mice 1 day before irradiation in dose 8.5 Gy	43.2 ± 2.1	0.4 ± 0.2
3.	Injection (i/v) into recipient mice 10^5 bone marrow cells from femur of intact donors	63.4 ± 3.2	12.5 ± 1.7
4.	Injection (i/v) into recipient mice 10^5 bone marrow cells from femur of donors mice treated 1 day before by Gastrofyline	74.5 ± 3.4*	19.8 ± 2.3 *

* - a significant difference with the corresponding control group /number 3/ ($p < 0.05$).

Table 3. Influence of Gastrofyline (25 ml/kg) administrated *per os* one day before preparation of bone marrow cells on the concentration of CFU-S (on 10^5 bone marrow cells) in bone marrow of mice with injection of cyclophosphamide (200 mg/kg)

№	Groups with different mouse bone marrow cells transplanted to irradiated in dose 8.5 Gy recipient-mice	Mean number of exogenous spleen colonies, (M±m)
1	Bone marrow cells from intact control donor-mice	12,3 ± 1,2
2	Bone marrow cells from donor-mice after *per os* administration of Gastrofyline 1 day before preparation of bone marrow	18,7 ± 2,0 *
3	Bone marrow cells from donor-mice with injection of cyclophosphamide 2 hours before bone marrow preparation	7, 2 ± 1,1 *
4	Bone marrow cells from donor-mice after *per os* administration of Gastrofyline 1 day before bone marrow preparation and with injection of cyclophosphamide into these donor-mice 2 hours before transplantation of bone marrow cells into recipient-mice	14,5 ± 2,1 **

* - a significant difference with the corresponding control group /number 3/ ($p < 0.05$).
** - a significant difference with the corresponding control group /number 3/ ($p < 0.05$).

Table 4. Action of Gastrofyline administrated *per os* in various concentration one day before local gamma-irradiation in dose 12 Gy on skin reactions of mice-hybrids F1(CBAxC57Bl/6), male, 3 months old, tested for maximal and mean skin reaction of hind leg in conventional units (9, 10). There were 12 animals in each group

№	Groups	Values of the maximal skin reaction in conventional units, M ± m	Values of the mean skin reaction in conventional units, M ± m
1.	Control, 12 Gy	2,4 ± 0,3	1,7 ± 0,2
2.	Gastrofyline, 40 ml/kg, one day before irradiation in dose 12Gy	1,5 ± 0,2 *	1,1 ± 0,2 *
3.	Gastrofyline, 25 ml/kg, one day before irradiation in dose 12Gy	2,0 ± 0,3	1,5 ± 0,2
4.	Gastrofyline, 5 ml/kg, one day before irradiation in dose 12Gy	2,3 ± 0,3	1,6 ± 0,2

- a significant difference with the control group ($p < 0.05$).

Table 5. Action of Gastrofyline administrated *per os* in concentration 40 mg/kg one day before local gamma-irradiation of tumor in dose 10 Gy on Lewis lung carcinoma growth and metastasizing on day 22th after tumor transplantation. There were 15 animals in each group

№	Groups	Mean tumor value on day 22th after transplantation of tumor, mm^3 (M±m),	Mean number of lung metastases on day 22th after transplantation of tumor, (M±m)
1	Control, nonirradiated tumors	564 ± 41	23.5 ± 3.8
2	Nonirradiated tumors from mice treated by Gastrofyline	550 ± 38	25.2 ± 4.6
3	Tumors irradiated in dose 10 Gy	232 ± 21	16.2 ± 3.0
4	Tumors irradiated in dose 10 Gy from mice treated by Gastrofyline	216 ± 17	14.9 ± 3.3

Table 6. Action of Gastrofyline administrated *per os* in concentration 40 mg/kg one day before onset of courses every days injection of cyclophosphamide (200 mg/kg in 4 daily injections on days 9-12th) on Lewis lung carcinoma growth and metastasizing on day 22th after tumor transplantation. There were 15 animals in each group

№	Groups	Mean tumor value on day 22th after transplantation of tumor, mm^3 (M±m)	Mean number of Lung metastases on day 22th after transplantation of tumor, (M±m)
1	Control mice with intact LLC	579 ± 45	21.8 ± 4.4
2	Mice with LLC treated by Gastrofyline on day 8th after tumor transplantation	567 ± 44	22.5 ± 4.9
3	Mice with LLC and injection of cyclophosphamide on days 9-12th after tumor transplantation	343 ± 29	12.3 ± 4.1
4	Mice with LLC treated by Gastrofyline on day 8th and injection of cyclophosphamide on days 9-12th after tumor transplantation	352 ± 28	11.4 ± 4.5

Conclusion

It was established, that GF is capable to decrease damage of normal tissues during chemotherapy or radiation therapy but without protective action on tumor growth and production of metastasizes, and that this biocomplexes stimulated hematopoietic stem cells to proliferate in vivo with increase its concentration in animal bone marrow. Absence of toxicity, disposable application *per os* does "Lutex" and "Gastrophyline" be perspective biocomplexes for application in modern practical medicine.

References

[1] T.N.Vorobeva ., Yu.I. Lubimov, V.A.Sushencov : Patent RU 2033797, (1992) (in Russian). .

[2] H. Ikeya., H., Kitagava: Patent US 5728391, (1998).

[3] E.G. Smirnova, Yu.I. Lubimov, T.G. Malinina, E.Yu.Lubimova: *Biochemistry J.,* 67 (11), 1539-1544 pp. (2002).(in Russian).

[4] I.Ye. Andrianova, T.G. Malinina, V.A. Glushcov, Yu.J. Lubimov: *Radiat. Biol. and Radioecology,* 44(4), 412-414 pp. (2004) (in Russian).

[5] J.E. TILL, E.A. Mcculloch : *Radiat. Res.,* 14(2), 213-222 pp. (1961).

[6] J.E. TILL, E.A. Mcculloch : *Ann. N-Y. Acad. Sci.,* 114(1), 115-125 pp. (1964).

[7] A.G. Konoplyannikov. *Radiobiology of stem cells, Energoatomizdat,* 120 pp. (1984) (in Russian).

[8] W.W. Smith, R.A.Budd, *J.Cornfild: Radiat. Res.,* 27(3), 363-368 pp. (1966)

[9] S.B. Field : *Radiology,* 93(4), 915-920 pp. (1968).

[10] O.K. Kurpeshev: *Diss. Thesis,* (1977) (in Russian).

[11] E.J. Ainsworth : *"Perspectives in radioprotection"* (Eds. Weiss J.F. and Simic M.G.), Pergamon Press, 223-242 pp. (1998).

In: Research Progress in Biotechnology
Editor: G. E. Zaikov, pp. 15-26

ISBN: 978-1-60456-000-8

Chapter III

Obtaining the Natural Food Additives from Fruit-Berry Raw Material

I.P. Mamleyeva, G.N. Dudikova and T.M. Orljuk
RSE «Research and production center for food processing industry»,
050060, Kazakhstan, Almaty, av. Gagarin, 238-a
Tel./fax: (3272) 49 89 42; E-mail: niipp@samal.kz

Abstract

Extraction parameters of fruit-berry raw material by liquid carbonic acid are perfected. Fruits of garden wild strawberry Chezena and Redgontlit grades, blackberry Mynai Shmyrjova grade, sea-buckthorn berries, dewberry without thorns, pressed skins of red grapes Almaly and Ilijskiy are selected as research objects.

The structure of fruit -berry raw material, CO2-extracts and fruit-berry CO2-raw material has been investigated. It is shown, that CO2-raw material is the valuable product due to the contents of vitamins C, E and ß-carotin alongside with CO2 extracts.

Keywords: Fruit -berry raw material, CO2-extracts, CO2-raw material, vitamins.

Introduction

Fruits, berries and grapes are the rich source of vitamins, mineral substances, carotenoids, phenolic compounds, organic acids, food fibres and other biologically valuable substances. Their value grows during last years when sharp deterioration of ecological conditions is marked, physical loadings of people, especially in the cities, are considerably reduced. Fruits and berries less energy-generating, in comparison with other sources of food. Their introduction in a diet promotes a dietary orientation of a meal and improves assimilation of animal origin meals.

In spite of the fact that in 2004-2005 years the average per capita consumption of fruits, berries and a grapes of own manufacture in Kazakhstan has increased up to 26-27 kg [1], it is necessary to take into account, that the use of fresh fruits and berries carries a seasonal nature, therefore requirement of biologically active substances of fruit, berry cultures and a grapes remains high. One of the ways maintaining the population such micronutrients is production of natural food additives on fruit-berry and grapes raw material basis.

Basing on chemical compound analysis, such cultures as currant, grapes, sea-buckthorn berries, mountain ash, wild strawberry, etc., are highly prospective for production of food additives such as dyes, aromatizes, antioxidants.

As a basic way of getting of food additives, CO_2- extractions have been chosen

Materials and Methods

Garden wild strawberry, black currant, sea-buckthorn berries, a grey blackberry (without thorns), red cultural grapes have been chosen as objects of research.

Garden Wild Strawberry - Fragaria Ananasa

Berries of wild strawberry contain 10-15 % of sugars, organic acids (salicylic, apple, cinchona, etc.) with prevalence of a citric acid, cellulose, pectinaceous and tannins, nitrogenous compounds, alkaloids, salts of iron, phosphorus, calcium, cobalt and manganese, vitamins of group B, ascorbic and folic acids, carotin, flavones [2-3].

Currant Black – Ribes Nigrum L

Berries contain up to 400 mg % of an ascorbic acid, up to 16 % of sugar, 4 % of organic acids (wine, amber, citric, salicylic, apple, nicotinic), pectinaceous and tannins, vitamins B_1, B_6, E, P, K, carotin, glucosides, radio oils, антоциан мальвин, phytoncides, major mineral elements and trace substances. The basic advantage of fruits is the low contents of the enzymes destroying an ascorbic acid owing to what they are a valuable source of vitamins [2-3].

Sea-Buckthorn Berries – Hippophae Rhamnoides L

The sea-buckthorn berries take a special place due to the contents of valuable biologically active substances in its fruits, bark and leaves. The contents of vitamin C varies from 150 up to 900 mg% and more. P-active connections of sea-buckthorn berries are submitted by riboflavinoids (100-200 mg%). From other water-soluble vitamins in sea-buckthorn berries are found thiamine (B_1), riboflavin (B_2) and a folic acid. The rich color scale of ripe fruits of sea-buckthorn berries (from various tints of yellow, orange up to brightly red) is connected to presence of carotinoids. Manoeuvrable sugar (up to 5 %), organic acids, irreplaceable amino acids, pectins, polyphenols [2-3] are contained in sea-buckthorn berries fruits alongside with vitamins.

Grey Dewberry (without Thorns) - Rubus Caesius

Fruits contain sugar (up to 3-5 %) - glucose, fructose, saccharose, xylose, ascorbic acid, carotin, vitamins B_1, B_2, E, nicotinic acid, organic acids (mainly apple, also oxalic, citric, salicylic), phenol carbolic acids and their derivatives, catechines (epicatechine), flavonoids, leukoanthocyanidins, anthocyanins, pectinaceous, tannic and aromatic substances, carotin. Plenty of potassium salts (up to 200 mg in 100 g), copper and manganese [2-3] is revealed in berries.

Grapes Cultural - Vitis Vinifera L

Fruits of grapes contain sugar (sometimes up to 20 % and more), cellulose, organic acids, ascorbic acid, vitamins of group B, pectinaceous substances, mineral substances and trace substances – potassium, calcium, magnesium, iron, manganese, cobalt, enzymes – invertase, pectinase, protease and lipase, tannins. The thin skin of berries contains tannic and painting substances [4]. Husks of red grapes are rich with antioxidants and melanin, possessing P-vitamin activity and high sorption ability in relation to ions of metals. Alongside with melanin the complex of phenolic compounds presented in husks possesses the directed radiation protection action.

Fruits of garden wild strawberry grades Chezena and Redgontlit, blackberry Mynai Shmyrjova grade, sea-buckthorn berries, dewberry without thorns, pressed skins of red grapes Almaly and Ilijskiy are selected as research objects.

Preparation and technological preparation of fruit-berry raw material is carried out on the standard methods [5].

Extraction of biologically active substances was carried out by method of undercritical CO_2-extraction of vegetative raw material. Working off the modes of extraction of fruit-berry raw material is carried out on the following parameters: temperature, °C; pressure, MPa; an output of a product, %.

Researches of biochemical structure of CO_2-extracts have been carried out on "GX" chromatograph of «Agilent Technologies» (USA) in helium stream at corresponding temperature modes, and gas chromatograph «Crystal Lux 4000 M» in a stream of nitrogen.

The contents of toxic elements in raw material have been defined by atomic - absorptive method on atomic absorber "Analyst".

Aromatic substances [ethers (volatile oils), aldehydes, aromatic spirits], vitamins C, E, ß-carotin have been defined by methods of HELC [7].

Definition of physical and chemical parameters of raw material, extracts and extraction cakes was carried out on the methods stated in [7].

Results

Technological Preparation of Raw Material

By the carried out researches it is revealed, that in Kazakhstan there is a sufficient volume raw-material of fruit-berry cultures and a grapes for working off and introduction in production of technology of reception of natural food additives [8,9].

Gathering of raw material carried out at the maximal accumulation of biologically active substances, i.e. during the achievement of maximal maturity by fruits and berries. At primary processing raw material the got impurities, defective fruits and berries have been deleted, fruit steams have been left if available.

As known, fruits and berries contain 85-90% of water. By preparation of raw material to extraction by carbonic acid, the dehydration modes of raw material by drying have been selected. To the maximal preservation of labile substances various kinds of dehydration applied: natural, without artificial heating; thermal, with artificial heating; sublimation.

Berries of wild strawberry have been dried up in two ways. In the first case berry have preliminary crushed in a blender and have dried up on sublimation laboratory dryer LP3 "Jouan". In the second case it has been dehydrated on thermal way in a drying case.

Sea-buckthorn berries, currant and blackberry with fruit steams have been dried in the whole kind in a drying case. For drying acceleration currant black was preliminary scalded in boiling water within 2-3 minutes.

The grapes have been taken as husks after separation of juice together with crests.

The data received after drying of fruits, are resulted in table 1.

The temperature mode in a drying case included two stages: 45-50°C (the basic drying) and further drying at higher temperature: 65-70°C. The currant alongside with vitamins contains volatile oils; therefore temperature of drying did not exceed 50-60°C.

The output of a dry product has made 7,2-48,7 % from initial quantity. The good output of a dry product is received from grape husks, black mountain ashes and sea-buckthorn berries, that is more effective for the further researches from the economic point of view.

Table 1. Parameters of drying and yield of raw material by preparation for process extraction

Product Name	Crude weight, g	Dry weight, g	Time of drying, h	Temperature, C°	Yield %
Grapes Almaly	3 000	1 462	68	63…80	48,7
Grapes Ilijsky	7 600	3 212	91	63…65	42,3
Sea-buckthorn berries	6 000	1 521	215	55…63	25,4
Blackberry	10 800	1 950	67	63…70	18,1
Currant	21 700	3 770	82	63…55	17,4
Wild strawberry by sublimation dryings	2 400	207	24	-	8,6
Wild strawberry of usual drying	2 600	188	89	63	7,2

Carrying out of Extraction Process

Moisture content in the dried up product is 7,6-11,7 %. The most effective for carbon-dioxide extraction structure of raw material from the point of view of the maximal surface mass-exchange and the best drainage properties is the petal thickness of 0,1-0,3 mm. But before to receive a petal, the raw material should be crushed. The dried up product have been crushed on the standard equipment – crusher KDU-2, leading up the size of particles of extracted raw material up to 0,5-1,5 mm.

The crushed raw material was loaded into an extractor of experimental installation and was filled with a liquid dioxide of carbon. Temperature of liquid carbonic acid limited with 22-25°C.

Reception of the CO_2-Extract of Wild Strawberry

Carrying out of continuous extraction process within 12 hours has led to reception of aromatic water fraction of 8 ml. Through 24 h - an additional output of aromatic water fraction of 4,5 ml, through 36 h – an additional output of 3 ml aromatic water fractions (table 2).

The general output for continuous extraction process duration of 36 hours has made 15,5 ml of water fraction or 7,7 % of volume of raw material. After 36 hours of continuous extraction experimental works have been stopped, not having achieved reception of pitch fraction. Taken extraction cakes possessed still a weak smell of a strawberry and residual humidity. It testified about non-extracted aroma, the possible reason – too big humidity of raw material.

Table 2. Parameters of extraction of berry and grapes raw material

Name of raw material	Working pressure in extractor, MPa	Extraction temparature, °C	Extraction duration, h	Extraction output, g (%)
Wild strawberry	5,7-6,5	22-25	12-36	15,5 (7,7)
Blackberries	6,2	23	6,0	8,0 (0,44%)
Currant	6,1	20	3,5	12,0 (0,42)
Grapes Ilijsky	6,2	23	6,0/48	12,0 (0,60)
Grapes Almaly	6,4	20	9,0/16	22,0 (1,50)
Sea-buckthorn berries	5,6-5,8	20	7,0/17	14,0 (0,90)

Reception of the CO_2-Extract of a Blackberry and Black Currant

For reception of the CO_2-extract from dry fruits of a blackberry and a black currant the dried up berries with residual humidity of 15-20 %, were crushed up to a flour of rough grinding condition. The amount of extracted raw material received from blackberry has made

1800 g, from black currant – 2860 g. Process of extraction of black currant included preliminary 15-hour infuse in liquid carbonic acid. Color and taste of extraction cakes of blackberries and a black currant practically has not changed and corresponded to initial raw material.

Reception of the CO_2-Extract of a Grapes and Sea-Buckthorn Berries

Dried up grape and berries of sea-buckthorn berries husks with residual humidity of 10 % have been crushed up to a condition of a rough grinding flour. The volume received of extraction raw material of a grapes of grade Almaly did not exceed 2/3 of volumes of an extractor (1520 g), a grapes of grade Ilijsly has sufficed on full volume of loading (2100 g), the volume of loading by sea-buckthorn berries has made 75 % from full volume of an extractor (1550 g). Modes of carrying out of process of extraction are resulted in table 2.

Research of Biochemical Structure of Fruit-Berry Raw Material and CO_2-Extracts

With the purpose of confirmation of expediency of the chosen fruit-berry raw material the content of its basic components is determined. Results of research are shown in table 3.

The content of water in fruits and berries is within the limits of 80,5 – 88,0 %. Except water the structure of fruits and berries includes carbohydrates, organic acids, vitamins, mineral and nitrogenous substances, fats, cellulose. The total of sugars changes in limits from 3 up to 15 % and depends on culture and conditions of cultivation. sugars are submitted by glucose, fructose and saccharose.

The basic component of cytoderms of a vegetative fabric is cellulose. Its content in the investigated fruits and berries is within the limits of 1,6 – 4,8 %. Albumens content in vegetative raw material is comparisonably low: 0,6 – 1,5 %. Complex ethers of glycerin and the higher carboxylic acids in researched raw material are presented in insignificant amount: 0,2 – 0,6 %, exception is made with sea-buckthorn berries: 5,4 %.

Huge value at processing fruits and berries has their acidity (contents of organic acids). Acids, basically, are presented by apple acid, also citric, oxalic, wine.

Mineral substances are presented as salts of organic and mineral acids, and also a component of some high-molecular compounds – vitamins, fibers, and enzymes. The general contents of mineral substances changes from 0,4 up to 1,5 %. The majority is presented by trace elements: potassium, calcium, magnesium, sodium, phosphorus, and iron. Potassium is most of them in berries.

Vitamin C is presented in fruits in plenties, especially in a currant and sea-buckthorn berries – 200 mg / 100 g. ß-carotin (carotinoids) paint fruits and berries in orange-red and yellow colors. The biggest content of this vitamin is marked in sea-buckthorn berries. In sea-buckthorn berries also there is a significant amount of fat-soluble vitamin E.

Table 3. Nutrition and biology value of plant raw material (for 100 g)

Name of raw material	Carbohydrates	Cellulose	Organic acids	Mineral substances						Vitamins		
				Na	K	Ca	Mg	P	Fe	β-carotine	C	E
	g			mg						mg		
Grapes Ilijsky	15,4	1,6	0,8	26	225	30	17	22	0,6	30	6	0
Blackber-ries grey (without thorns)	4,4	2,9	2,0	21	208	30	29	32	1,0	100	15	1,6
Wild strawberries Chezena	7,5	2,2	1,3	18	161	40	18	23	1,2	30	60	0,8
Sea-buckthorn berries	5,7	2,0	2,0	4	193	22	30	9	1,4	1500	200	18
Currant grades Minay Shmyrjev	7,3	4,8	2,3	32	350	36	31	33	1,3	100	200	0,9

Quality of initial raw material is characterized also by the contents of heavy metals, which should not exceed allowable concentration. In this connection corresponding researches which results are shown in table 4.

Table 4. Heavy metals content in initial raw material

Sample	Cd, mg/kg	Pb, mg/kg	Zn, mg/kg	Cr, mg/kg	Cu, mg/kg	Ni, mg/kg
Garden wild strawberries	0,00102	0,0085	0,1173	0,12	3,88	0,33
Blackberries grey (without thorns)	0,00042	0,0062	0,918	1,10	6,11	0,22
Currant	0,00041	0,0075	1,740	0,01	4,58	0,30
Grapes "Almaly"	0,00029	0,0077	1,364	0,73	9,24	0,07
Grapes "Ilijsky"	0,00025	0,0061	1,478	0,98	7,45	0,04

Apparently from the data resulted in the table, the content of heavy metals, except copper, is stacked in the norms established by SanPiN 2.3.2.560-96 in Republic Kazakhstan. The contents of copper exceeds allowable norm (5 mg / kg) in a blackberry and a grapes.

Further the vitamin structure of CO_2-extracts and the CO_2-raw material which has stayed after extraction (figures 1, 2) has been investigated

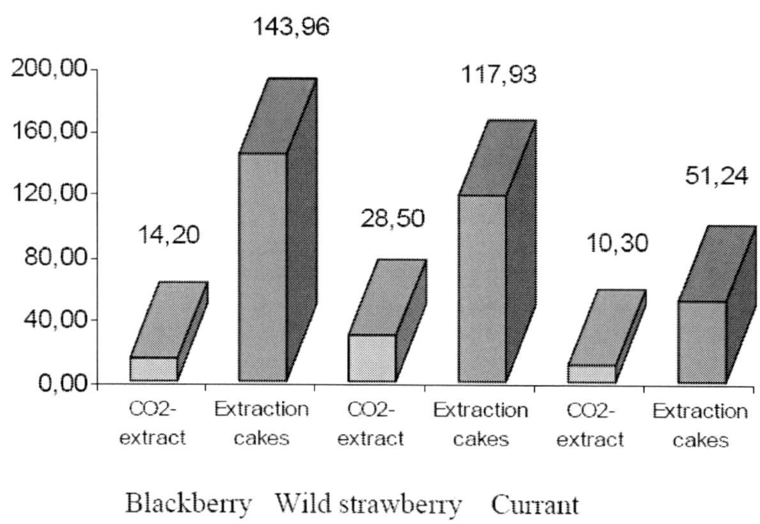

Figure 1. Vitamin C content in CO_2-extraction products, mg/100 g.

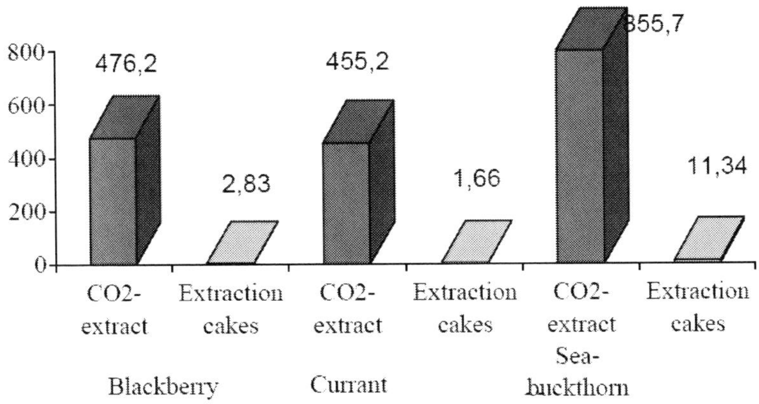

Figure 2. Vitamin E content in CO_2-extraction products, mg/100g.

Apparently from the received data, the vitamin complex is distributed between the CO_2-extract and the CO_2-raw material remaining after carrying out of extraction process. The data will be coordinated to results of definition of the contents of vitamins in raw material in view of humidity of products.

Taking into account that fact, that liquid carbonic acid in a combination to pressure possess the bactericidal action, the received CO_2-extracts and CO_2-raw material do not contain microorganisms and meet the requirements of industrial sterility that is positive at storage of these additives.

The total amount of flavonoids in grape husks and in CO_2-raw material remained after extraction (figure 3) has been determined.

CO_2 extraction poorly influences at flavonoids extraction– their basic content is in fruit CO_2-raw material, remaining as a by-product of extraction process.

Fruit-berry extraction cakes, received after carrying out CO_2-extraction, has kept many properties of initial raw material and had very attractive kind.

The put experiences have shown, that it can be used for reception of natural dyes.

The similar data are received for a black currant and Aronia melanocarpa.

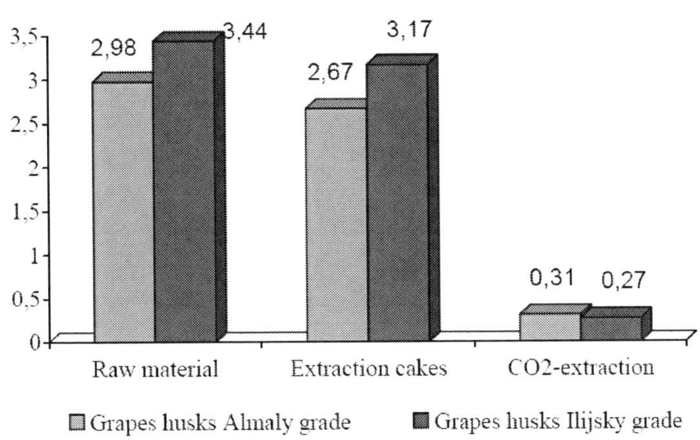

Figure 3. Flavonoids content in plant raw material and CO_2-extraction products from grapes husks, % DC.

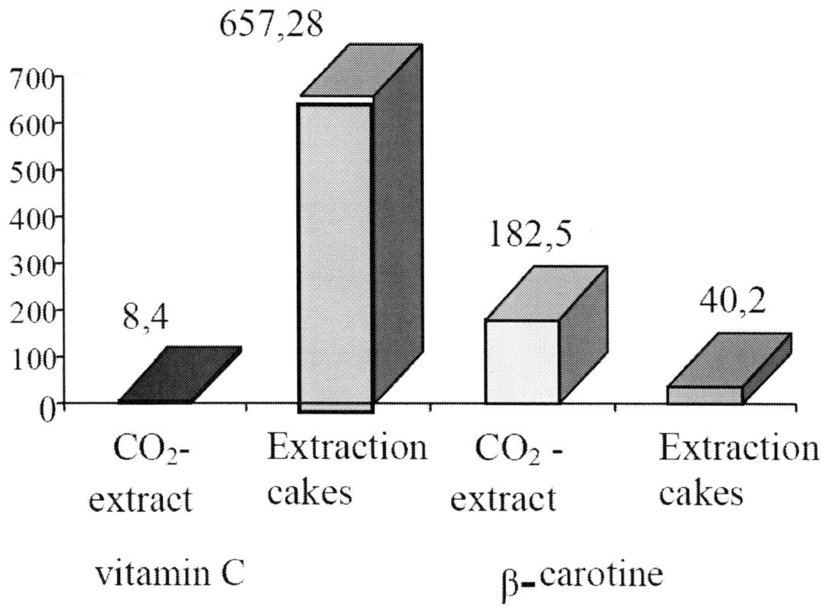

Figure 4. Vitamin content in CO_2–extraction products from sea-buckthorn , mg/100g.

Discussion

By virtue of lines of the objective reasons vitamin-mineral deficiency [10] is observed at total country population.

The wide set of natural components of food is necessary for normal functioning of an organism of the person and all his systems (micronutrients) to which the organism is genetically adapted, and at least, the balanced vitamin-mineral complexes.

Plants are the basic source of bioflavonoids, vitamins, glycosides, alkaloids, trace elements, organic acids and other natural substances regulating and stimulating physiological functions of an organism.

CO_2-extaction is a new, arising industry in Kazakhstan, capable to provide pharmacological and food manufactures with products.

As a rule, objects of CO_2-extraction are volatile oils and aromatic herbs [11].

Working off and reception of CO_2-extracts from fruit-berry raw material allows to expand assortment of domestic, natural, non-polluting food additives, as sources micronutrients, at the Kazakhstan market.

To the complex compounds taken from vegetative raw material by liquid dioxide of carbon, concern aromatic, flavouring, cereous and resinous substances, in part lipoids and group of biologically active connections where fat-soluble vitamins and provitamins enter among others.

As our researches have shown, a chemical compound of the fruit-berry raw material selected for carrying out CO_2-extraction is submitted by complex connections. Fibers, carbohydrates, mineral and vitamin complexes, tannins, organic acids concern to them lipids, etc. Checked up on the contents of heavy metals the raw material answered the norms established by Sanitary rules and norms 2.3.2.560-96 in Republic Kazakhstan. In blackberry and grapes the contents of copper is a little bit overestimated. However for preventive and the medical products, the enriched children's meal consumption of copper up to 5 mg day is norm as copper concerns to necessary trace elements.

Thus, the chosen fruit and berry cultures are valuable raw material for use in manufacture of CO_2-extracts. Thus it is necessary to note, that manufacture commercially CO_2-extracts and CO_2-raw material in the country is practically provided with an initial raw material.

At the first stage of work we determined the contents of the basic vitamins in CO_2-raw material and CO_2-extracts. The results received are showing, that the contents of vitamin C, is concentrated basically in fruit-berry CO_2-raw material. Its content in sea-buckthorn berries is 657,28 mg / 100g, good parameters at grapes Ilijsky grade (280,6), blackberries (143,96) and wild strawberry (117,93) is especially high. Such parameter is explained by the high content of dry substances in final products. In CO_2-extracts of a currant and sea-buckthorn berries the ascorbic acid is kept in small amounts (10,5 and 8,4 mg / 100g accordingly), in both grades of grapes it is absent in general.

Vitamin E, by virtue of its nature (fat-soluble), is extracted directly into CO_2-extract of a blackberry, currant, sea-buckthorn berries in sufficient quantities (from 455 up to 855 mg / 100g). In grapes and wild strawberry vitamin E in extraction products is not defined, since its contents there is the lowest.

ß-carotin in a significant amount is present only at sea-buckthorn berries and is extracted by carbonic acid into CO_2-extract (182,5 mg / 100g).

Thus, due to the contents of vitamins C, E and ß-carotin, both the CO_2-extract, and the CO_2-raw material, stayed as a by-product at carrying out of process of CO_2 extraction are valuable vitamin products.

Presence of vitamins C and E, and also ß-carotin in extracts and extraction cakes is the important parameter for their use in the food-processing industry..

In the conclusion it is necessary to note, that we carried out preliminary experiences which have shown, that extraction cakes from the used fruit-berry raw material at carrying out CO_2-extraction can be used for reception of natural dyes on a basis of ethanol. Painting substances of a black currant, blackberry, grapes concern to group anthocyan dyes, yellow color of sea-buckthorn berries is caused by the high contents of other natural dyes – carotinoids. Received extraction cakes can also be used as natural aromatizer for giving to foodstuff corresponding taste and aroma.

References

[1] Brief statistical year-book of Kazakhstan. Almaty., 2005 (in Russian).

[2] G.L Pilat., A.A Ivanov: Moscow, 2002, 710 P.(in Russian).

[3] S.Y. Sokolov: *Medical information agency,* Moscow, 2000, 976 p. (in Russian).

[4] A.I Gozhenko, N.G Slavina., G.P. Ovchinnikov., A.A. Lazutkina : *Conference materials.* Moscow.-Leningrad, 1993, P.44-46. (in Russian).

[5] S.G. Ilchenko, A.T. Markh, A.F. Fan-Ung: *Food Industry,* Moscow, 1974, 423 p. (in Russian).

[6] Gost 30178-96 Raw materials and food products. *Atomic absorptive method of toxic elements determination.* (in Russian).

[7] The guide on methods of quality and safety of food products analysis: Under red of Skurikhin I.M., Tuteljan V.A. : *«Brandes», Medicine* ,Moscow,., 1998, 340 p. (in Russian).

[8] I.P. Mamleyeva : *Vestnik SPC Processing and Food Industry,* Alma-Ata, 2006, 3: 8-14. (in Russian).

[9] I.P Mamleyeva., A.S Aitimov.: *Health and disease,* Alma-Ata 2005, 3: 81-84. (in Russian).

[10] Kazakhstan: *National investigation of nutritional status of population,* 1996//UN/VOZ/KIP. – Almaty, 2001. (in Russian).

[11] B.M. Proskurin Food Ingredients. *Rae materials and additives.*2002, 1: 3-6. (in Russian).

In: Research Progress in Biotechnology
Editor: G. E. Zaikov, pp.27-34
ISBN: 978-1-60456-000-8

Chapter IV

Influence of Some Factors on Cell Free Synthesis of Cobalamines by Propionibacteria

L.I. Mingalieva, D.M. Ruzieva, F.A. Karimova and T.G. Gulyamova
Institute of microbiology, 7b A.Kadyri Street, Tashkent, Uzbekistan, 100128
e-mail: imbasru@uzsci.net; fax: 998-71-1442582

Abstract

It was established that isolated cytosol of strain *Propionibacterium acidipropionici 30* preserves cobalamine synthesizing ability of native cells. Qualitative composition of buffer solutions, duration and conditions of incubation for cobalamines synthesis in cell free system by cytosol of Propionibacteria cells were determined. It was established that optimum synthesis of cobalamines by bacterial cytosol requires anaerobic conditions. It was determined that increase of cobalamines yield by 30% takes place at addition of aminolevulinic acid at concentration 15 mg/l. Use of HPLC analysis allowed to establish that spectrum of produced metabolites at cell free synthesis is more narrow compared to metabolites spectrum synthesized by native cells.

Keywords: Propionibacteria, cobalamines, cell free synthesis

Cobalamines are complex tetrapyrrole compounds the mechanism of which formation is extremely complex and influenced by many factors limiting velocity of formation, their quantitative and qualitative content [1, 2, 3].

Currently, studies of biosynthesis pathways and possibility of regulation of mechanisms of cobalamines formation are conducted in cell free systems to eliminate influence of different factors. Numerous studies revealed that the most profitable method of cobalamines receipt is controlled microbiological synthesis with use of highly active producers [4, 5]. Propionibacteria are one of the well-known producers: metabolism of natural strains is primordially tuned for high level of cobalamine synthesis [6]. At the same time, Propionibacteria accumulate cobalamines intercellularly and do not secrete them in to environment. But, cultivation of Propionibacteria demands on observance of certain conditions: anaerobiosis, maintenance of medium pH, step-by-step addition of precursors. All these factors greatly affect bacterial growth velocity. However, establishment and maintenance of above mentioned conditions and following extraction of cobalamines from bacterial biomass are quite complex and labor-intensive procedures.

In these regards, cytosol fraction of bacteria was used to study regulation of cobalamines synthesis in local strain *Propionibacterium acidipropionici 30.* It allowed to avoid influence of above mentioned limiting factors of cultivation on synthesizing activity of the culture.

Methods

Studies were conducted on cytoplasmic fraction of cells of local strain *Propionibacterium acidipropionici 30* isolated from local dairy products at the laboratory of biochemistry and biotechnology of physiologically active compounds of the Institute of microbiology.

To obtain cytosol Propionibacteria were cultivated 3 days in anaerobic conditions at 30°C on agar and liquid nutrient medium of the following composition (g/l): corn steep – 20; glucose – 10; K_2HPO_4 – 2; $(NH_4)_2SO_4$ – 5; $CoCl_2$ – 0,01; agar 20; pH – 6.8-7.0. Following buffer solutions (pH 7.0) were used as incubation medium: №1 – 0.05M Na-phosphate; №2 – 0.02M Tris-HCl; №3 – 0.02M Tris-HCl, 0.25M sucrose, 3mM $MgCl_2$, 15mM KCl.

Biomass of culture cultivated on liquid medium was collected by centrifugation at 6,000 rpm during 30 min, twice washed out with buffer and frozen in the same buffer. Biomass of culture cultivated on agar medium was collected by buffer wash-out.

After repeated cycles of freezing-thawing cell suspension was homogenized in Potter's homogenizer. Obtained homogenate was centrifuged at 12,000 rpm during 30 min. Supernatant was cytoplasmic fraction which was used as enzymatic preparation in modeling regulation of cobalamines biosynthesis. Protein content was determined by Lowry method [6].

Obtained cytosol was incubated at room temperature, 30 and 37°C during 1, 2 and 3 hours with a addition of biosynthesis precursor aminolevulinic acid (ALA) at concentration from 5 to 25 mg/l to study influence of composition of buffer solutions, their pH, temperature factor, incubation duration on cell free synthesis. Total cobalamines content was determined on spectrophotometer. To do so, cyanidation of samples of incubated cytosol was conducted by addition of acidulous water pH 4.5-5.0 with 0.1% NaCN. Mixture was heated up 80°C during 20 min on water bath and after cooling its pH was adjusted to 4.5-5.0 and following optical density measurements were done at wavelength 548 nm.

Results and Discussion

Cell's cytosol fraction of strain *Propionibacterium acidipropionici 30* was obtained and used as enzymatic preparation to study cell free synthesis of cobalamines. Initially it was established that cultivation conditions influence on enzymatic activity of cytosol. Thus, it appeared that cytosol obtained from biomass of bacteria cultivated on agar medium synthesizes more vitamin per 1 mg of protein compared to cytosol obtained from biomass of bacteria cultivated on liquid medium.

Study of influence of qualitative composition of buffer solutions revealed that the most active synthesis takes place in buffer №3. Cobalamines yield in this buffer exceeded cobalamines yield at incubation in buffer №1 in 3-5 times (table 1). The time of incubation was also important factor. According to obtained data the highest cobalamine content in all tested buffers was observed on 2nd hour of incubation.

It is known that synthesis of vitamin B12 is more susceptible to strong influence of medium pH level than growth of Propionibacteria [7]. Our results conform to results of other scientists and testify that the highest vitamin quantity is synthesized at neutral pH of incubating medium (figure 1).

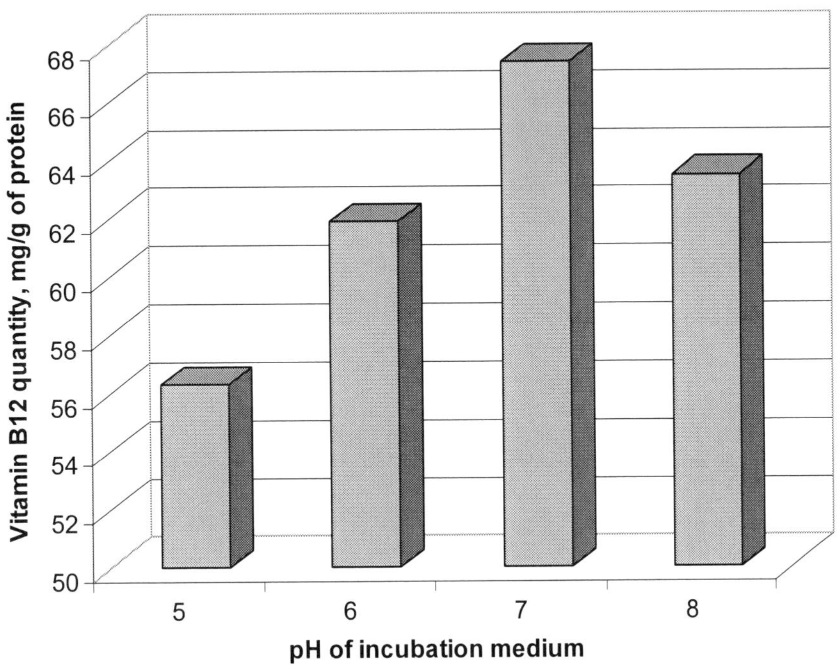

Figure 1. Dependence of cell free synthesis of vitamin B12 from pH of incubation medium.

Table 1. Dependence of vitamin B12 synthesis level in cell free system from qualitative composition of buffer solutions and incubation time

Buffer solution (pH 7.0)		Time of incubation, h		
		1	2	3
		Vitamin B12, mg/g of protein		
№ 1	0.05M Na-phosphate	33.1	34.4	31.1
№ 2	0.02M Tris-HCl	64.8	68.8	51.5
№ 3	0.02M Tris-HCl, 0.25M sucrose, 3mM $MgCl_2$, 15mM KCl	89.7	169.3	143

Enzymes vary on their resistance towards temperature. Study of temperature influence on cobalamines synthesis by cytosol fraction revealed that at 37°C the increase of total cobalamines content by more than 40% is observed compared to its content at room temperature (figure 2). Thus, it was established that for optimum cell free synthesis of cobalamines the cytosol incubation should be conducted in buffer №3 during 2 hours at 37°C.

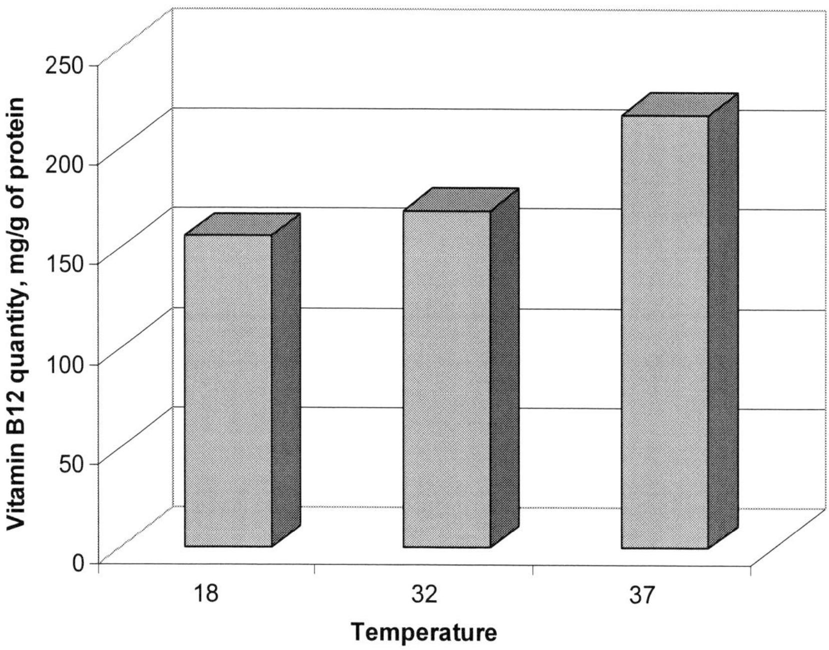

Figure 2. Dependence of cell free synthesis of vitamin B12 from incubation temperature.

Vitamin B12 biosynthesis includes numerous stages with participation of different enzymes. Biosynthesis pathway starts from formation of general precursor 5-ALA, which in different biological objects is synthesized by one of the following methods: via condensation of glycine with succinil-CoA or synthesis from glutamine acid [8, 9, 10]. This process is catalyzed by enzyme ALA-synthetase, which enzymatic activity affected by air oxygen [11, 12]. Thus, e.g., ALA-synthetase activity is higher in cell of *Micrococcus denitrificans* cultivated in anaerobic conditions.

In these regards it was interesting to study influence of oxygen on cell free synthesis of vitamin B12 by cytosol fraction of strain *Propionibacterium acidipropionici 30* to reach this goal culture fermentation and subsequent cytosol incubation was conducted in anaerobic and aerobic conditions in different variations (figure 3). It was established that optimum synthesis of cobalamines by cytosol of Propionibacteria is observed in anaerobic conditions. It is proved by fact that according to obtained data the level of cobalamines production in cell free system in aerobic conditions considerably decreases efficiency of synthesis, without regard to conditions in which the culture was cultivated itself. At the same time, enzymatic activity of cytosol obtained from culture cultivated in anaerobic conditions is twice higher than in culture cultivated in aerobic conditions.

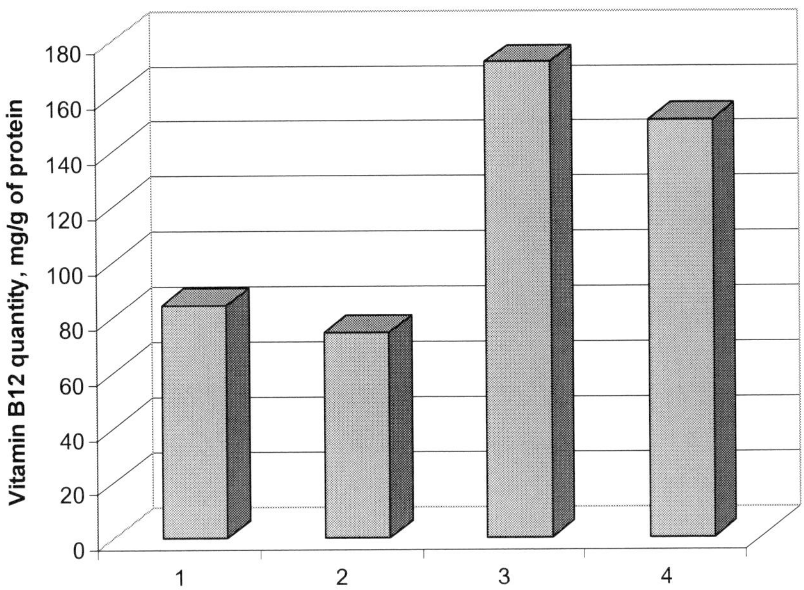

Figure 3. Dependence of cell free synthesis of vitamin B12 from oxygen availability. 1 – incubation of cytosol of aerobically cultivated cells at oxygen presence; 2 – incubation of cytosol of aerobically cultivated cells at oxygen absence; 3 – incubation of cytosol of anaerobically cultivated cells at oxygen absence; 4 – incubation of cytosol of anaerobically cultivated cells at oxygen presence.

Obtained data may testify that, apparently, the activation of key enzymes of synthesis of metabolites of tetrapyrrole line takes place in process of anaerobic cultivation of Propionibacteria and, thus, initial enzymatic activity is considerably higher in cell free system of anaerobic cultivated cells.

According to available data, activity of such important and studied by now enzymes like ALA-synthetase, activating process of production of ALA, the general precursor of all tetrapyrrole compounds, and cobalt-chelatase, incorporating cobalt ions in molecule of vitamin B12, is in direct dependence to availability of oxygen [13, 14, 15]. At that activity of both enzymes is greater in anaerobic conditions. Our results fully conform to available literature data.

Currently there are many different data available on 5-ALA synthesis by Propionibacteria but clear and reliable data are missing yet. Thus, recent studies of Japanese scientists revealed that ALA formation in *P.acidipropionici* occurs from glutamine acid and in *P.freudenreicii* from glutamate and glycine [16, 17]. Apparently, vitamin synthesis pathway is determined by Propionibacteria specie, availability of necessary substrates and other factors. In these regards we studied influence of glutamate and glycine as substrates for cell free synthesis of 5-ALA at vitamin B12 formation (figure 4). It was established that glycine and glutamate as substrates ambiguously influence the cell free synthesis of vitamin B12. Introduction of glycine increases synthesis of vitamin by more than 20% whereas introduction of glutamate has no effect at all.

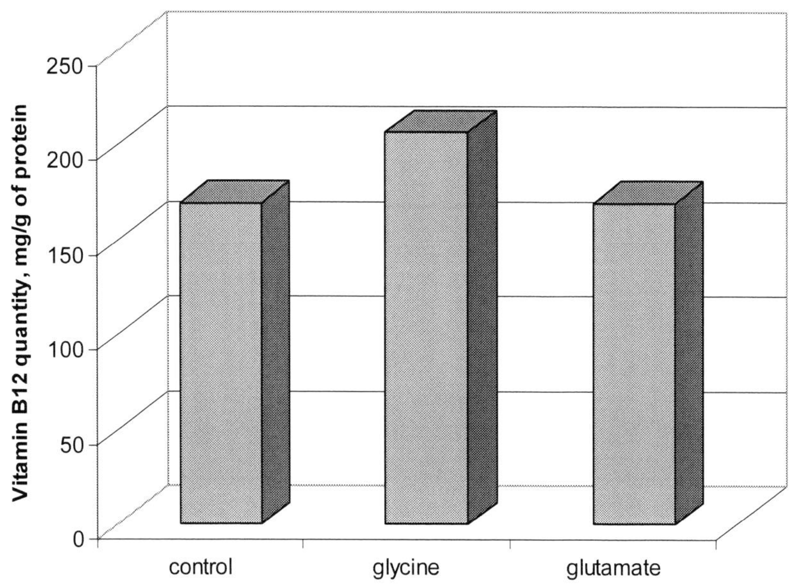

Figure 4. Dependence of cell free synthesis of vitamin B12 from substrate (concentration 20 mg/l).

According to data obtained by Moretti et al., apart of 5-ALA gamma-aminobutyric acid (GABA) is formed from glutamate as result of decarboxylation. GABA has general transport proteins with ALA. At the same time, 5-ALA has no own transport system [18, 19, 20, 21]. In these regards, it maybe suggested that at addition of exogenous glutamate, probably, not only increase of ALA synthesis takes place, but also of GABA. As result of the competition for proteins-carriers transportation of 5-ALA is decreased and this influences further level of produced vitamin B12, whereas exogenous glycine increasing only formation of ALA does not effect GABA production.

Considering obtained data we studied ability of cytosol fraction of Propionibacteria directly digest exogenous ALA at cell free synthesis of cobalamines. Results of conducted studies revealed that at presence of exogenous ALA increase of cobalamines accumulation in incubation medium takes place (figure 5). The optimum concentration of precursor is 15 mg/l with cobalamines yield increase by almost 30%. Further increase of ALA concentration results in decreasing vitamin B12 synthesis.

Aiming identification of cobalamines spectrum produced in cellular and cell free systems of Propionibacteria the qualitative analysis of metabolites was conducted on HPLC. It was established that among metabolites of cell free synthesis only two derivates are present (cyancobalamine and dicyancobinamide) whereas native cells also produce adenosylcobalamine (table 2). The initial cytosol at cell free synthesis contains only cyancobalamine and addition of exogenous ALA leads to appearance of minor fraction of dicyancobinamide. Whereas, addition of ALA in cellular system changes only quantitative ratio of different cobalamine forms and qualitative composition remains same.

Comparison of metabolites content in extractions of native cells and in extractions of cytosol makes clear that at cell free synthesis the spectrum of metabolites is narrow, which maybe considered as an advantage. Thus, obtained data reveal that isolated cytosol of Propionibacteria preserves relatively high cobalamine synthesizing ability which suggests practical availability of the method of cell free synthesis of vitamin B12.

Table 2. Qualitative composition of products of cellular and cell free synthesis of strain *Propionibacterium acidipropionici 30* at presence of ALA

ALA concentration, mg/l	CN-Cbl, γ/ml	$(CN)_2$-Cbd, γ/ml	Ado-Cbl, γ/ml	CH_3-Cbl, γ/ml
Extraction of cellular cytosol (control)	2.7	-	-	-
25	1.8	0.12	-	-
15	3.6	-	-	-
10	-	0.04	-	-
Extraction of native cells (control)	0.75	0.24	0.16	
25	1.5	0.2	0.12	-
15	3.3	0.04	0.06	-
10	0.3	-	0.28	-

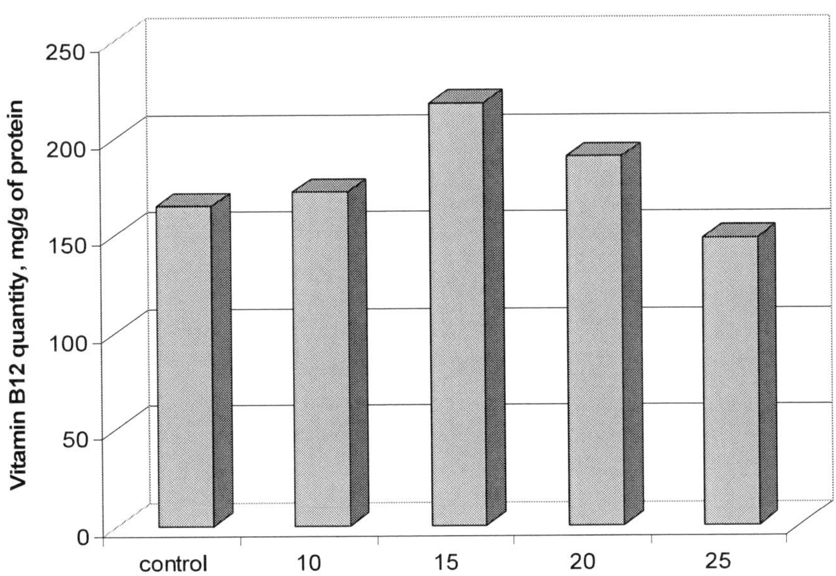

Figure 5. Dependence of cell free synthesis of vitamin B12 from ALA.

References

[1] V.Y.Bykhovsky, N.I.Zaitseva, A.A.Eliseev: *Appl. Biochem. Microbiol.,* 34, 1-18 (1998).

[2] A.I.Scott, C.A.Roessner: *Biochem. Soc. Trans.,* 30, 613-620 (2001).

[3] M.J.Warren, E.Raux, H.L.Schubert, J.C.Escalante-Semerena: *Nat. Prod. Rep.,* 19, 390-412 (2002).

[4] E.Raux, H.L.Schubert, M.J.Warren: *Cell. Mol. Life Sci.,* 57, 1880-1893 (2000).

[5] V.Y.Bykhovsky, N.I.Zaitseva: *Itogi nauki i tehniki,* 32, 173 p. (1989). (in Russian).

[6] Practical chemistry of protein. Ed. A.Darbre, *Mir,* Moscow, 1989, p. 622. (in Russian).

[7] L.I.Vorobyova: *MSU publishing house,* Moscow, 1995, 288 p. (in Russian).

[8] A.Ano, H.Funahashi, K.Nakano, Y.Nishizawa: *J. Biosci. Bioeng.,* 88, 57-60 (1999).

[9] P.M.Jordan: McGraw-Hill, New York, 1990, 55-122.

[10] J.Moser, W.D.Schubert, D.W.Heinz, D.Jahn: *Biochem. Soc. Trans.,* 30, 579-584 (2002).

[11] A.R.Battersby: Wiley/VCH, *Weinheim,* 1998, 47-61.

[12] A.I.Scott, C.A.Roessner, P.J.Santander: Wiley, New York, 1999. 537-556.

[13] D.Heldt, A.D.Lawrence, M.Lindenmeyer, E.Deery, P.Heathcote, S.E.Rigby, M.J.Warren: *Biochem. Soc. Trans.,* 33, 815-819 (2005).

[14] K.Iida, M.Kajiwara: *Biochem.,* 39, 3666-3670 (2000).

[15] N.Gardner, C.P.Champagne: *J. Appl. Microbiol.,* 99, 1236-1245 (2005).

[16] R.D.Willows, V.Lake, T.H.Beale: *J. Bacteriol.,* 185, 3249-3258 (2003).

[17] K.Murakami, Y.Hashimoto, Y.Murooka: *Appl. Environ. Microbiol.,* 59, 347-350 (1993).

[18] C.A.Roessner, K.H.Huang, M.J.Warren, E.Raux, A.I.Scott: *Microbiology,* 148, 1845-1853 (2002).

[19] M.Bermudez Moretti, S.R.Correa Garcia, M.S.Chianelli, E.H.Ramos, J.R.Mattoon, A.Batlle: Int. J. Biochem. *Cell Biol.,* 27, 169-173 (1995).

[20] M.Bermudez Moretti, S.R.Correa Garcia, A.Batlle: *Biochem. Biophys. Res. Commun.,* 272, 946-950 (2000).

[21] S.R.Correa Garcia, M.Bermudez Moretti, E.H.Ramos, A.Batlle: *J. Biochem. Cell Biol.,* 29, 1097-1101 (1997).

In: Research Progress in Biotechnology
Editor: G. E. Zaikov, pp. 35-40
ISBN: 978-1-60456-000-8

Chapter V

Utilization of Supercritical Fluids in Enzymatic Conversion of Cotton Biomass

G.A. Muratov[1] **and C. Kim***[2]*
[1] Institute of Microbiology of AS RUz, Tashkent, Uzbekistan
[2] Division of Chemical Engineering and Biotechnology, College of Engineering, Ajou University, Suwon 442-749, South Korea

Abstract

The purpose of this study is to identify the types of cotton biomass that can be used to produce sugars and evaluate utilization of supercritical fluids in enzymatic hydrolysis of cotton fibres, as well as optimization of whole process. The study includes: 1) enzymatic hydrolysis of unriped cotton fibes from green bolls (cultivar "Tashkent-6" and "Omad"of Gossypium hirsutum L.); 2) application of commercial cellulase enzymes from different sources: Trichoderma viride, Trichoderma reesei and Aspergillus niger; 3) identification and description of glucose yields after enzymatic hydrolysis; 4) comparison of enzymatic hydrolysis kinetics at supercritical and atmospheric conditions; 5) optimization of process conditions in enzymatic hydrolysis of cotton fibers; 6) suggestions for the logical next steps.

Keywords: cellulase, cotton fiber, glucose, supercritical fluids, sort "Tashkent-6", sort "Omad", Gossypium hirsutum L., Trichoderma viride, Trichoderma reesei, Aspergillus niger.

* G.A. Muratov; Tashkent, Uzbekistan, 700128, Kadyriy str., 7B; E-mail: gairatm@hotmail.com, Fax: (998)71-1427129
* C. Kim; E-mail: ckim@ajou.ac.kr, Fax: (81082)-31-214-8918

Introduction

Utilization of cellulose materials is known to produce great amounts of wastes whose processing can yield usable products, including carbohydrates. Experimental studies were carried out on evaluation of potential for producing glucose from cotton wastes in carbohydrate conversion. Cotton processing industry wastes (such as a cotton lint, linter, bur, stalks, cottonseed hulls, chiganak, cotton dust, linter waste of oil works) have been found to be little prone to hydrolysis due to their highly crystalline state or the high content of lignin in some of them [1]. For this reason, we choose kind of raw material not investigated previously: cotton fibers from bolls of early autumn picking (20, 30, 40 day's age). This matter is discarded unprocessed, although it is rich in cellulose (75-80%). Note that cellulose fibers are amorphous at this age, and their enzymatic degradation requires no pretreatment [2].

The activities of cellulase enzymes from Trichoderma viride, Trichoderma reesei and Aspergillus niger used in enzymatic hydrolysis were maintained in atmospheric condition and in the supercritical carbon dioxide solvent. Various sorts of cotton plant were used. As it was shown in our previous research, particularly unriped fibres differ the greater contents of amorphous cellulose in them, than crystalline, with their maximal degree of conversion to glucose at enzymatic hydrolysis. Besides 20-40-days-old fibres are form waste products of cotton-growing at the autumn-winter collecting. It is known, that supercritical fluids serve as a particularly interesting class of solvents, which may be used for enzyme-catalyzed reactions. Supercritical fluids are materials at temperature and pressure above their critical points. Above the critical point, molecular thermal energy exceeds the attractive forces between molecules, and a gas-like and liquid-like state exist; consequently, the properties of supercritical fluids bridge the gap between the properties of liquids and gases. Biochemical relations in supercritical fluids were demonstrated first in 1985 by Randolph et al. [3]. The enzymatic hydrolysis reaction in supercritical carbon dioxide to produce glucose from cellulose was investigated in 1996 by Zheng [4]. Substantial improvement in the yield was possible in supercritical medium as compared to that obtained in the hydrolysis carried out in atmospheric condition [5,6]. The purpose of the present research was an application of supercritical carbon dioxide to enzymatic hydrolysis of cotton boll fibres (sort Tashkent-6 of Gossypium hirsutum L.) by commercially available cellulases. Experiments carried out at pressures from 100 up to 200 атм., temperatures 30-80^{0}C and time of hydrolysis from 2 till 60 hours.

Materials and Methods

For our experiments various cotton sorts Tashkent-6, Omad, 149-F, Termez-31, C-6037, Ashhabad-25 of species G.hirsutum L., G.barbadense L., G.arboreum L. and G.herbaceum L. maintained at the Laboratory of cytology and genetics of Research Institute of Cotton Breeding and Seed Production under Scientific Production Center of Agriculture of the Ministry of Agriculture and Water Management of Uzbekistan were used. We choosed 30-

day-old fibers of sort Tashkent-6 for a comparative study of enzymatic hydrolysis of cotton fibers in a supercritical carbon dioxide and atmosphere conditions.

For enzymatic conversion of cotton fibers cellulase complexes from Trichoderma viride (6.9 units/mg), Trichoderma reesei (10.4 units/mg) and Aspergillus niger (1.18 units/mg) were used. These enzymes were purchased from Sigma Chemical Co. , USA (one unit liberate 1.0 mmole of glucose from cellulose in one hr at pH 5.0 and at 37°C).

The experiments were performed in shaking incubator (Korea Manhattan Co.) with 15-20 ml temperature-controlled (50°C) cells, speed control (100 rpm.) at pH 5.0 and 0.1 M acetate buffer and in the batch reactor system, shown in figure 1.

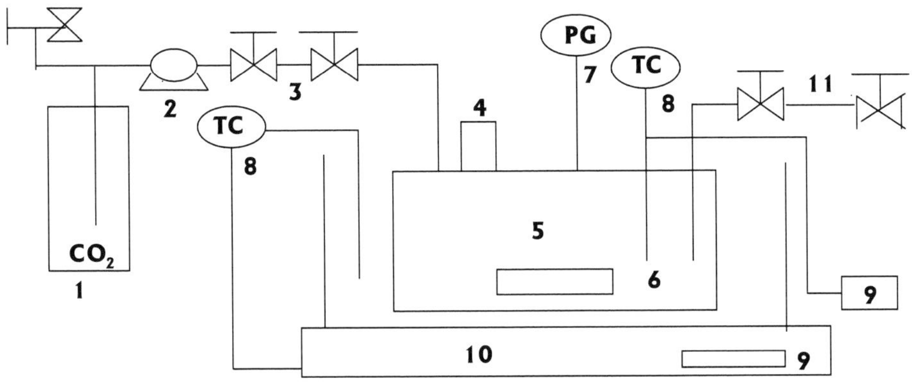

1.CO_2 bomb; 2. Liquid pump; 3. CO_2 input line ; 4. Back pressure regulator ; 5. Reactor; 6. Magnetic bar; 7. Pressure indicator ; 8. Thermocouple; 9. Temperature indicator ; 10. Digital stirring bath ; 11. Sampling valve.

Figure 1. Schematic diagram of apparatus.

Reaction vessel (250 ml working volume) made of SS 316 and designed to sustain the pressure of up to 250 atm, was placed in a digital water stirring bath equipped with a temperature controller. The constant temperature inside the vessel was maintained within the accuracy +0.5°C. The reaction vessel was filled with 0.1 M acetate buffer (pH 5.0) and flushed out air with carbon dioxide (99.99% of purity) from a dip-tube container. The reaction vessel was enclosed and carbon dioxide was immediately pumped into the reaction vessel using a LDC Analytical Minipump (Gearmotor, USA) until the desired pressure was reached and the pressure was maintained with a back pressure regulator (Tescom Co, USA). The reaction mixture was agitated by a magnetic stirrer. After a designated length of time kept at a certain pressure and temperature, the sample was immediatly taken out through the sampling line and then analyzed to measure the glucose concentration. The stability of enzymes in supercritical carbon dioxide was determined by incubating an enzyme in each supercritical condition. The relative enzyme activity was measured at optimum condition of atmospheric pressure.

The concentration of glucose in the solution was determined by glucose oxidase-peroxidase method [3] and a glucose analyser (YSI 1500, USA). Optical density was measured in a spectrophotometr (Shimadzu UV-1201, Japan). Glucose yields were assayed at 480 nm. The concentrations were: substrate, 2%; enzyme, 0.6%. Temperature: 50°C.

Results

In this study the hydrolysis of cotton cellulose by different cellulases was performed in supercritical carbon dioxide. Cellulase enzymes process cellolose under certain conditions, cleave glucoside bonds, reduce the degree of cellulose polymerization, and eventually cleave it to glucose. Experiments on stability of cellulases were carried out in supercritical carbon dioxide. The activity of cellulase was found sustained at the pressure of up to 160 atm. No change of activity was observed at the temperature of up to 50°C, which was an optimum temperature for the cellulase activity at ambient pressure. The activity, however, decreased down to 50% at 70°C and the pressure of 200 atm. The cellulase activity was mostly maintained at the pressure of 120 atm for 48 hours and 50°C. This observation is also in accordance with the experimental results [7]: in the hydrolysis of cellulose the glucose yield showed its maximum at 50°C, the temperature at which activity of cellulase was the highest. Figure 2 show results of comparative hydrolysis of cotton sort "Omad" fibers submitted under optimum conditions by cellulase enzymes from Trichoderma viride, Trichoderma reesei and Aspergillus niger during 48 hours. Maximal yield of glucose is observed at 48 hour hydrolysis of 30-days-old fibres by cellulase from Trichoderma viride and makes 15.5 mg/ml. These data also prove to be true calculations on a degree of conversion of an enzyme-substrate complex of 20, 30 and 40-days-old cotton fibres on weight (table 1). Thus the degree of conversion varied from 33.9 % (A. niger) up to 92.3 % (T. viride). And only in a case cellulase enzyme from Trihoderma viride the degree of conversion about 90 % was observed. Table 2 summarizes some results of experiments on comparative 48 hour enzymatic hydrolysis of 30-days-old cotton fibres of cultivar Tashkent-6 performed in supercritical and atmospheric conditions. Despite of significant deviations at use various enzymes at different pressures, the yields of glucose in supercritical carbon dioxide almost always is higher, than at atmospheric condition. Distinctions in a glucose yields depending on used enzymes speak differences in structure of enzymes of cellulase complex and activities of its each separate component. The maximal yields of glucose was observed at use cellulases from Trichoderma viride during 48 hours at supercritical condition (120 atm, 50°C) and has made 13.9 mg/ml.

Table 1. Degree of conversion after 48-hourly hydrolysis of "Omad" sort cotton fibers different days of maturation by cellulases of various origin

Days of maturation	A. niger	T. reesei	T. viride
20	53.9%	90%	92.3%
30	33.9%	80.8%	91.5%
40	38.4%	77.8%	87.7%

Table 2. Glucose yields (мg/мl) of 30 days-old cotton fibers of sort Tashkent-6 (G.hirsutum L.) in supercritical and atmospheric conditions by cellulases from Aspergillus niger and Trichoderma viride; 20 мg/мl of substrate, 41 units/ml of cellulase, 100 rpm, pH 5.0, and 50°C

Enzyme reaction	Hours of hydrolysis	A. niger	T. viride	T.viride + A.niger
Atmosphere	6	3.0	7.0	6.4
	12	4.1	8.6	7.2
	24	5.4	9.8	9.5
	36	5.9	10.6	10.9
	48	6.0	11.0	10.6
Pressure		160 atm	120 atm	100 atm
Supercritical CO_2	6	3.3	8.2	8.5
	12	4.4	11.2	9.5
	24	6.4	12.6	9.9
	36	7.1	13.5	10.7
	48	7.2	13.9	10.7

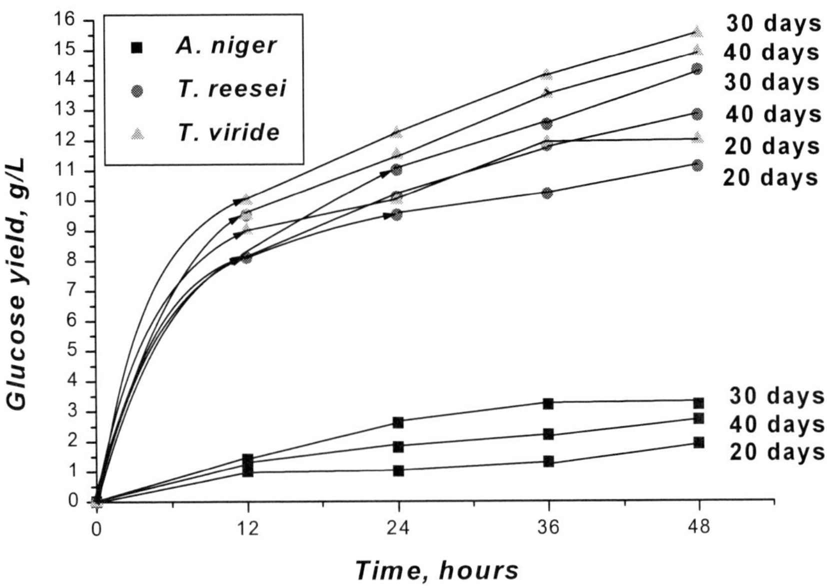

Figure 2. Comparative 48-hour hydrolysis of "Omad" cultivar cotton fibers by cellulases of various origin; 20 mg/ml of substrate, 6 mg/ml of cellulase, 100 rpm, pH 5.0, and 50°C.

Discussion

In the present experiment, an effort was made to improve the current methods of cellulose biodegradation and study of possibility of using cellulases from different sources on their activity toward cotton fibres of different species as well as conducting of enzymatic hydrolysis of cotton-growing wastes in supercrtical fluids in compare to atmosheric condition. All our experiments performed up to now have thrown some lights on several aspects of this process, namely, (i) optimal condition for conducting this experiment determined: pressure 120 atm at 50°C for 48 hours, (ii) highest activity was showed by cellulase enzymes from T.viride (degree of conversion up to 92.3%), (iii) best substrate is 30-days old cotton fibres from Tashkent-6 (G.hirsutum L.), although there is no big difference in compare to fibres from other cultivars and species, (iv) the total productivity of glucose in supercritical carbon dioxide was about 1,2 times of that obtained at atmospheric pressure.

These results indicate that the method of supercritical carbon dioxide can be considered further as one of potential ways of bioconversion of wastes of cotton growing. But at the same time the given problem demands further more detailed studying depending on concrete conditions of realization of the hydrolysis, used enzymes and substrates.

References

[1] A. P. Sinitsyn, A. A. Klysov, M. L. Rabinovich, A. B. Gusakov, A. M. Morozov: *Itogi Nauki Tekh.,* Ser.: Biotechnologiya, Moscow: Vses. Inst. Nauchn.-Tekh. Inf. 12: 3-151, 1988 (in russian).

[2] G. A. Muratov, M. M. Rakhimov, B. T. Yuldashev: *Appl. Biochem. Microbiol.* 34 (5): 477-479, 1998.

[3] T. W. Randolph, H. W. Blanch, J. M. Prausnitz, C. R. Wilke: *Biotechnol. Lett.* 7: 325-328, 1985.

[4] Y. Zheng, G. T. Tsao: *Biotechnol. Lett.* 18 (4): 451-454, 1996.

[5] C. Kim, K. D. Oh: *Korean J. Chem.Eng.* 4: 105-112, 1987.

[6] H. S. Lee, W. G. Lee, S. W. Park, H. Lee, H. N. Chang: *Biotechnol. Tech.* 7 : 267-270, 1993.

[7] C. Y. Park, Y. W. Ryu, C. Kim: *Korean J. Chem. Eng.* 18 (4): 475-478, 2001.

In: Research Progress in Biotechnology
Editor: G. E. Zaikov, pp. 41-51
ISBN: 978-1-60456-000-8

Chapter VI

Bioremediation of Chronic Black Mineral Oil Pollution and Railway Sludge with the Oil-Oxidizing Preparation "Rhoder" (Field Tests)

***V.P. Murygina*[*1], *A.I. Tumasyants* [1], *M.I. Iankevich* [2], *D.V. Zhukov* [1] *and S.V. Kalyuzhnyi*[1]**

[1] Department of Chemical Enzymology, Chemistry Faculty,
Moscow State University, Leninskye gory 1-11, 119992, Moscow, Russia
[2] Joint Stock Co., LTD "ILMA-EKO", quay of Bypass channel,
150, 190020, St.-Petersburg, Russia

Abstract

During summer of 2006, the field tests were performed in St.-Petersburg on bioremediation of atypical pollutants (chronic black mineral oil pollution in a ground and railway sludge) using the oil-oxidizing preparation "Rhoder". Total hydrocarbon content (THC) in chronic black mineral oil pollution in soil varied from 115 to 120 g/kg of dry material (DM) and in the railway sludge – from 58.8 to 77.0g/kg DM. Black mineral oil, diesel, ashes, residues of coal and other organic substances polluted the sludge. Triple treatment by the preparation "Rhoder" was applied to the above mentioned pollutions. A phytoremediation with seeds of oats and Timothy grass was conducted simultaneously with the third treatment by the "Rhoder". As a result, the THC decreased by 85-90% and 54 % in the chronic black mineral oil pollution and in the railway sludge, respectively.

Keywords: pollution, black mineral oil, railway sludge, bioremediation, oil degrading bacterial strains, preparation "Rhoder", phytoremediation.

[*] e-mail vpm@enzyme.chem.msu.ru, tel: +7 (495) 939-50-83; Fax: +7 (495) 939-54-17

Introduction

Restoration of the territories polluted by oil hydrocarbons (HC) is quite relevant for Russia nowadays [1] because it is practically impossible to avoid pollution of soils, water surfaces, underground waters, sediments, as oil still remains a basic energy source on the Earth. The most attractive technologies of restoration HC polluted territories (both from economic and ecological points of view) are bioremediation technologies including an introduction of HC degrading preparations.

One of the promising domestic bacterial oil degrading preparation is the "Rhoder" developed by us earlier [2]. The preparation consists of two bacterial strains belonged to the genus *Rhodococcus* and isolated from soil polluted by crude oil. These bacterial strains were proven to be non toxic, non pathogenic and non toxigenic for warm-blood animals and deposited into the All-Russian Collection of Microorganisms. The Phylogenetic position of "Rhoder" strains was determined on the basis of 16s RNA sequences. An immunochemical method of identification for both strains was developed and can be used for their detection during bioremediation trials. Finally, the Sanitary-and-Epidemiologic Certificates (No77.99.24.929D.005978.10.04 and No77.99.03.003D.003020.06.05 on liquid and dry forms of the "Rhoder", respectively) were obtained for application of the preparation in the field conditions.

A working solution of the "Rhoder" is prepared by dissolving the preparation in water (0,1-1,0 kg of dry powder per 1000 m^3 of water.) up to the necessary concentration of HC oxidizing cells in the solution ($1*10^6$ – $1*10^7$ cells/ml). Mineral components (sources of nitrogen, phosphorus and potassium in concentration of 0.02 % - 0.2 %) are added to the working solution before application for bioremediation. Usually the preparation is multiply applied on the polluted site with an interval of 2-3 weeks.

The advantage of the preparation "Rhoder" is an ability of its HC-degrading bacteria to attach firmly to the inter-phase of oil-water owing to their lipophilic cellular walls. Bacterial strains of the "Rhoder" possess of synergism at joint degradation of oil HC. Antagonistic relations with indigenous microorganisms are not observed. The "Rhoder" and products of HC biodegradation do not render any harmful effect on the environment. Bacterial strains of the "Rhoder" loose their activity in a case of HC absence as sources of nutrition and energy, and they are consumed by the Protozoa and other elementary organisms in water and soil and fill up humus layer of soil. The temperature and pH ranges of the "Rhoder" action are 10-32^0C and 5.8-7.8, respectively. The preparation "Rhoder" operates in a wide range concentration of oil pollution (till > 20%) but it is most effective at a low level of pollution (0.1-0.5 %), which remains after application of mechanical or chemical restoration methods.

Numerous testing of the preparation "Rhoder" in laboratory, pilot and field-scale conditions in Moscow region, Western Siberia, Komi Republic (1995-2002) has shown it high efficiency for bioremediation of different hydrocarbon contaminated soils as well as water surfaces: 57-99% removal efficiency of HC from water surfaces and 29-73 % removal efficiency of oil contamination from wetland and soils after twice or triple application of the "Rhoder" depending on initial HC concentrations and weather conditions [3-6]. In 2004, the preparation "Rhoder" (the dry form) was examined on bioremediation of oily sludge in

China, province Shandong, and showed a decrease of HC contamination by 45-56 % after three applications with the initial THC of 120-250 g/kg DM [7].

In 2006, according to the Government Contract No 02.447.11.3001, pilot and field tests of the "Rhoder" (the dry form) were conducted on atypical substrata: chronic black mineral oil pollution of the soil on the territory of the State Obukhov factory and railway sludge nearby the railway station "Ruchji" (both St.-Petersburg). This paper reports the results of these tests.

Materials and Methods

Pilot Tests

Two sites (size - 5m*5m) on the territory of the State Obukhov factory polluted by black mineral oil (chronic pollution) have been dug up on the depth of 15-20 cm and fenced by tapes. The ground on the sites was powder-like with a lot of stones. One site was visually more polluted by viscous black mineral oil than the other one. The pollution of the sites was heterogeneous and exceeded 100 g/kg DM. The preliminary inspection of HC pollution of the sites was executed in the end of June 2006. In a month, sand in quantity of 50 and 30 kg per site (for the first site and the second one, respectively), fertilizer (7.5 kg/site) and lime were added to the ground of the sites.

The sites were once again dug over and solution of the "Rhoder" (250 g in 200 L of technical water) was spread with watering cans on a surface of each site. The ground then was loosened by a rake. Three treatments with the "Rhoder" were made within an interval of 2-3 weeks with simultaneous digging up of sites. Fertilizer and lime (dolomite) (7.5 kg/sites) were added once again before the second treatment of the ground with the "Rhoder" because the ground was very poor with biogenic elements. During the third treatment with the "Rhoder", the same fertilizer was added in quantity of 0.2 % in a working solution of the preparation. Simultaneously with the third treatment of the ground with the "Rhoder" a mix of oats and Timothy-grasses seeds (about 1 kg on each site) were manually sowed onto the sites. All seeds were preliminary treated by the anti-stress agent. The samples of ground for analyses were taken using a 5-dot scheme from the upper layer (15.0 cm of depth) before each treatment with the "Rhoder" and 1.0 month later after the last treatment.

Field Experiment

This experiment was performed on a cement platform (size 20m*40m*0,5m) containing 20 tons of railway sludge removed from railway embankments and inter-rail space. The sludge was contaminated with diesel, black mineral oil, ashes, residues of coal and other organic compounds. This sludge was mixed with 40 tons of old sludge remained after bioremediation (in 2005) with a preparation "Ruden" immobilized on a peat. For our experiment, it was allocated 30 tons of mixed sludge with the THC of 58.8–77.0 g/kg DM.

50 kg of fertilizer (azophoska) and 49 kg of lime were manually put onto a surface of 30 tons of railway sludge, ploughed up by a mini-tractor and loosened by a rake. The working solution of preparation "Rhoder" (2.5 kg of the dry form per 1 ton of technical water) was watered on 30 tons of sludge and then this sludge was loosened with a rake. The most probable number (MPN) of active hydrocarbon oxidizing (HCO) cells in the working solution was about of $1*10^6$ cells/ml. Before the third treatment with the "Rhoder", 50 kg of the same fertilizer and 49 kg of lime were put additionally to the sludge because pH of the railway sludge decreased from 7.1 up to 6.3, and nitrogen and phosphorus were exhausted by 86 and 87 % (on the average), respectively. Simultaneously with the third treatment with the "Rhoder", about 10 kg of seeds mix of oats and the Timothy grasses (preliminary treated with the anti-stress agent) were manually sowed on the sludge.

The samples of the railway sludge for analyses were taken from 6 dots along the length of cement platform before each application of the "Rhoder" and also in a month after start of the phytoremediation.

Weather Conditions

Bioremediation on the territory of the State Obukhov factory and the railway station "Ruchji" started in the end of July 2006. The weather during experiments was mainly dry (rains practically were absent), warm (the temperature of air varied from 20 to 28°C). In September, the temperature of air became lower (14-20°C) and short-term rains occurred. The ground on the experimental sites at the State Obukhov factory and railway sludge on the railway station "Ruchji" were not loosen and not watering between the "Rhoder" applications.

Analyses

The samples of ground and railway sludge were analyzed on the THC, mobile forms of nitrogen and phosphorus, pH and microorganisms. The THC was determined by gravimetric method, whereas column chromatography was used for fraction analysis of HC [8]. Mobile forms of nitrogen ($N\text{-}NH_4^+$) and phosphorus ($P\text{-}PO_4^{3-}$) were determined with colorimetric method [9].

Microorganisms (ammonifying, nitrifying, oligonitrophylic, HCO ones) were determined using elective media (Meat-Peptone agar (MPA), Chapek-Dox agar, Ashby agar, liquid Raymond media with oil, respectively) and classical microbiological method determination of MPN. Inoculated Petri dishes and test tubes were incubated under 29°C within 3-5 and 14 days, respectively. The MPN of the HCO strains in the "Rhoder" were determined in the same way using liquid Raymond media with oil and MPA.

Results and Discussion

Pilot Tests on the Territory of the State Obukhov Factory

The results of the chemical analysis of the ground polluted by chronic black mineral oil showed that the THC amounted for 120 and 115 g/kg DM on the first and second sites, correspondingly. After modifying of the polluted ground with sand, the THC decreased to 42.4-66.7 g/kg DM. The triple application of the preparation "Rhoder" followed by phytoremediation further decreased the THC by 85% and 90 % (on the average) for the first and second sites, respectively (figure 1).

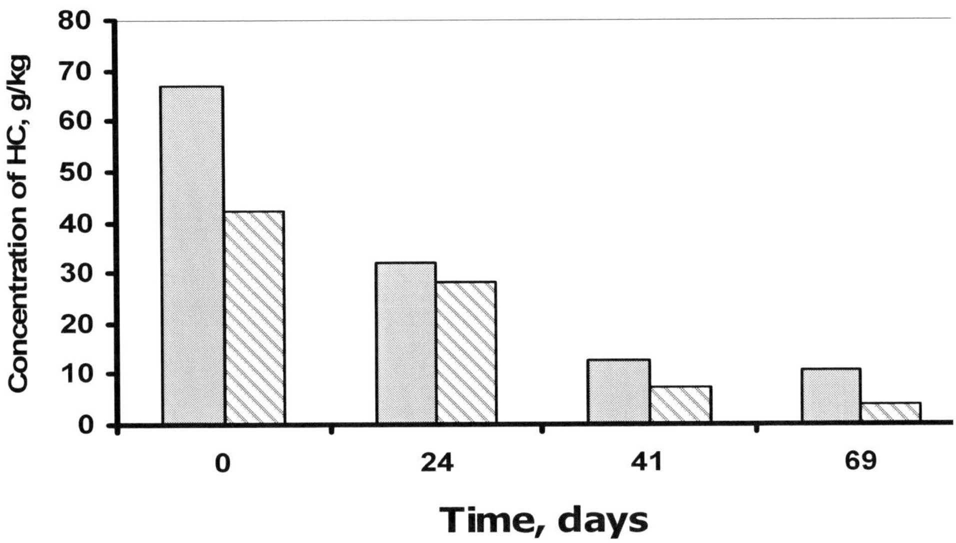

Figure 1. Degradation of HC in chronic black mineral oil pollution during bioremediation with the "Rhoder" in two sites on the territory of the State Obukhov factory, 2006.

The HC fraction analysis of initial chronic black mineral oil pollution showed essential differences in the content of aliphatic and asphalt-resinous HC on the sites (figure 2). The concentration of aliphatic HC on the first site was about 13 g/kg, whereas on the second one – in 2 times higher. On the contrary, the asphalt-resinous part of HC on the first site was about 27 g/kg, i.e., higher than on the second one (about 18 g/kg). For the aromatic HC, insignificant differences were not observed on both sites (figure 2.). In spite of these initial differences, the preparation "Rhoder" in combination with phytoremediation essentially decreased the concentrations of aliphatic, aromatic and asphalt HC by 91-95 % on both sites.

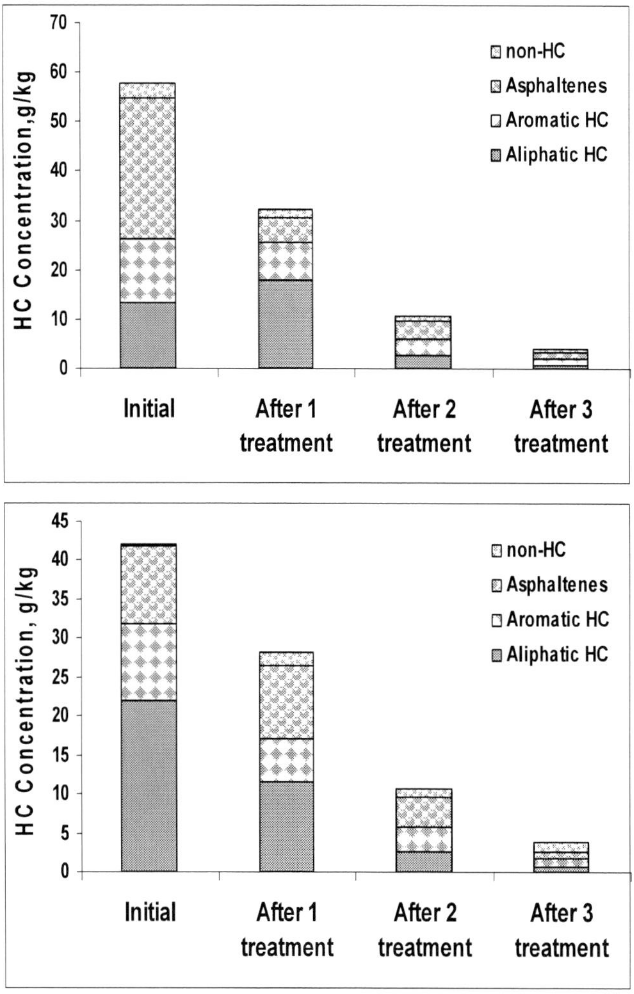

Figure 2. Changes of group compounds of HC in chronic black mineral oil pollution during bioremediation with the "Rhoder" (the State Obukhov factory, 2006.

Since pollution was old, a microbiological control of basic groups of microorganisms, responsible for biological activity of soil, revealed a presence of HCO, ammonifying, nitrifying and oligonitrophylic bacteria in rather high MPN (figure 3). The "Rhoder" introduction exerted not only an increase of number HCO microorganisms but also stimulated a growth of other groups of microorganisms. It is also seen that after consumption a bulk of HC in chronic black mineral oil pollution, the MPN of HCO cells decreased. The number of other groups of microorganisms also returned back to the initial MPN in autumn (figure 3). However, despite of initial rather high number of indigenous microorganisms in the ground of treated sites, an efficiency of the "Rhoder" application is obvious, since it allowed to decrease essentially the THC of chronic black mineral oil pollution in one (though rather warm) season and to stimulate a biological activity of the ground (figure 3).

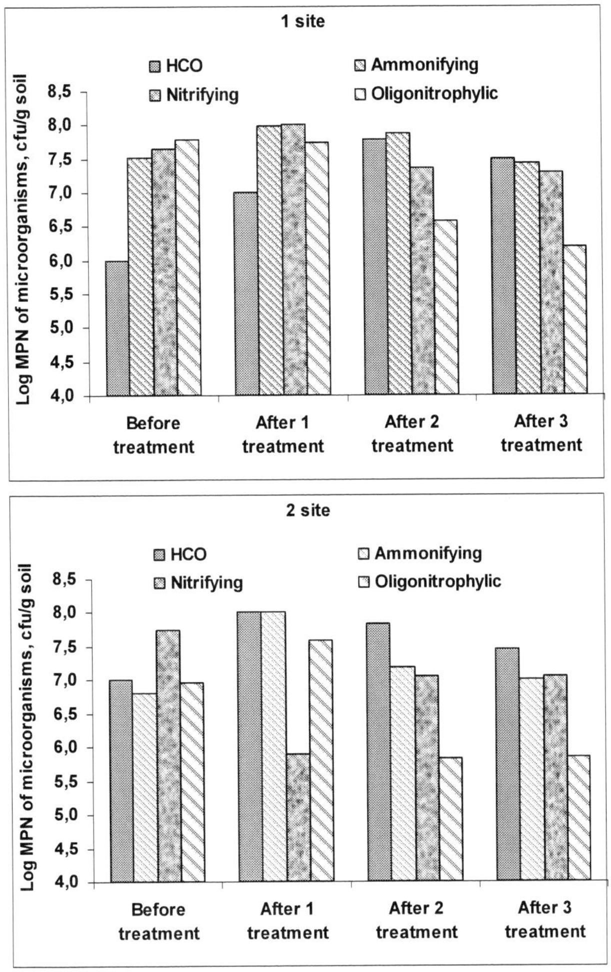

Figure 3. Monitoring of MPN of different groups of microorganisms responsible on the biological activity of ground polluted with chronic black mineral oil during bioremediation with the "Rhoder" on the sites (the State Obukhov factory), 2006.

Field Experiment at the Railway Station "Ruchji"

The analyses of the mixed (modified) railway sludge (30 tons) showed that it contained a relatively high number of HCO bacteria and other groups of microorganisms as well as that the average content of nitrogen and phosphorus in ground was low (table 1).

The THC in the modified railway sludge before bioremediation varied from 58.8 to 77.0 g/kg DM. The pollution was represented by aliphatic HC, aromatic HC, asphaltenes and non-extractable substance at the average content of 56.1, 21.7, 15.6 and 6.1 %, respectively (table 2).

Table 1. Agrochemical and microbiological characteristics of modified railway sludge before bioremediation (the railway station "Ruchji")

Samples	pH	$N\text{-}NH_4^+$, mg/100g	$N\text{-}NO_3^-$, mg100g	$P\text{-}PO_4^{3-}$ mg/100g	HCO cells/g	Ammo-nifiers., cfu/g	Nitrifiers, cfu/g	Oligo-nitrophiles, cfu/g
(1-2)	6.8	4.3	0.048	7.6	$9.0*10^7$	$1.2*10^8$	$9.5*10^7$	$1.0*10^8$
(3-4)	7.34	4.0	0.029	3.5	$9.0*10^7$	$2.3*10^8$	$4.5*10^7$	$5.3*10^7$
(5-6)	6.9	6.5	0.053	6.1	$9.0*10^7$	$2.3*10^8$	$4.5*10^7$	$5.3*10^7$

Table 2. Chemical characteristic of pollution of the railway sludge before bioremediation (the railway station "Ruchji")

Samples	Moisture %	THC g/kg DM	Aliphatic HC, %	Aromatic, HC, %	Asphaltenes, %	Non extractable, %
(1-2)	11.5	58.8	53.2	23.3	15.4	8.1
(3-4)	9.8	77.0	56.8	22.5	12.8	7.9
(5-6)	14.3	65.2	59.9	19.4	18.5	2.2

The triple application of the "Rhoder" with the subsequent phytoremediation decreased the THC content on the average by 54 % in the modified railway sludge (figure 4). During bioremediation, the concentration of aliphatic HC decreased by 90 %, aromatic HC - by 50 %, however, asphaltenes were practically not degraded (figure 5).

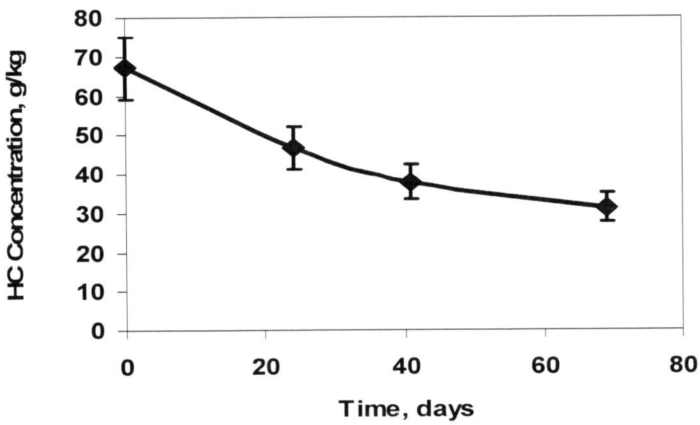

Figure 4. Degradation of HC in the railway sludge during bioremediation with the "Rhoder" (the railway station 'Ruchji"), 2006.

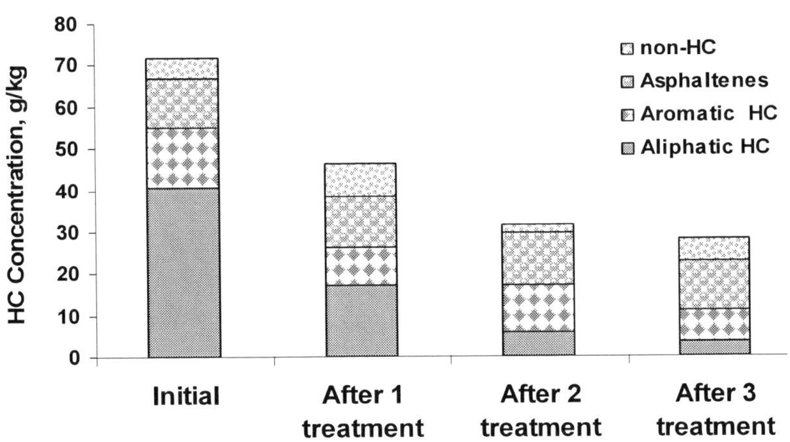

Figure 5. Changes of group compounds of HC in railway sludge during bioremediation with the "Rhoder" (the platform nearby the railway station "Ruchji"), 2006.

The MPN of HCO microorganisms in the railway sludge before bioremediation was rather high ($1*10^7$ cells/g DM of sludge). During bioremediation this group of microorganisms increased their number for 1-2 orders of magnitude and kept at a high level. Analysis of other groups of microorganisms describing a biological activity of the sludge showed that its bio-cenosis has already been developed, because the "Rhoder" was applied after bioremediation of the part of railway sludge by another oil-degrading preparation. Nevertheless, throughout the process of sludge decontamination, the number of these groups of microorganisms further increased for a while (figure 6). Introduction of HCO–bacteria with the "Rhoder" did not disturb a balance of developed microbial cenosis, moreover, promoted its prosperity (figure 6).

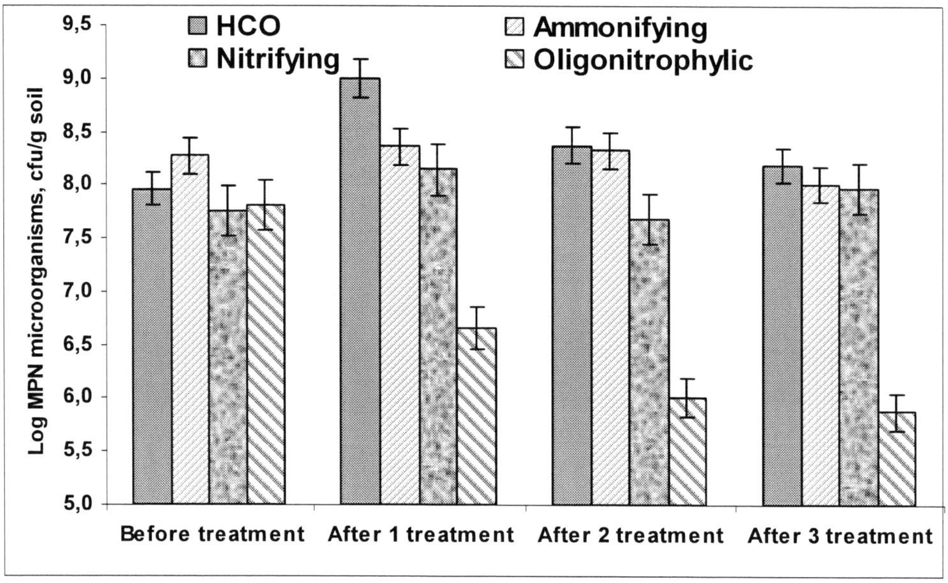

Figure 6. Monitoring of MPN of different groups of microorganisms responsible on the biological activity of the railway sludge during bioremediation with the "Rhoder" on the platform nearby the railway station "Ruchji", 2006.

Conclusion

The oil-oxidizing preparation "Rhoder" for one warm season (three applications) with the subsequent phytoremediation was capable to decrease the THC of chronic black mineral oil pollution of ground and railway sludge by 85-90% and by 54 %, respectively. The introduction of HCO-bacteria with the "Rhoder" into polluted ground or sludge did not disturb a balance of developed microbial cenosis.

References

[1] New Technologies for Clean up Oil Contaminated Waters and Soils, remaking and use Oil-Slimes. *Proc. Intern. Confer.* by ICS and UNIDO, 10-11 of December, Moscow, 27 (2001) (in Russian).

[2] V. P. Murygina, N. E. VoishvillO, S. V. Kalyuzhnyi. Patent RU No 2174496, 2001

[3] V. P. Murygina, M. Arinbasarov, S. Kalyuzhnyi *Biodegradation.* 11(6) (2000) 385-389.

[4] K. Salem, G. G. Mikailov, V. P. Murygina, I. V. Perminova, N. U. Grechicheva, S. V. Mecheriakov. in Nova Science Publishers., 2004, 29-39.

[5] V.P. Murygina, M.Y. Markarova, S.V. Kalyuzhnyi. *Environment International,* 31 (2005) 163-166.

[6] V. P. Murygina, S. V. Kalyuzhnyi, P. Gally, F. Stirly, Liu Hong, Ouyang Wei. in Nova Science Publisher, Inc. 2006, pp. 87-91V. Murygina, Liu-Hong, Ouyang-Wei, Yu-Yongyong, Xiu-Zengde, S. Kalyuzhnyi. in Proc. of 8 Intern. *Bioremed. Conference,* Greece. 3-7 June, 2006

[8] E. Bogomolov (Ed.) Modern methods of analysis of oil. Higher School, Moscow, 1984, 341 p. (in Russian).

[9] V. G. Mineev (Ed.) *Practical work on agro-chemistry.* MSU, Moscow, 2001, 688 p. (in Russian).

In: Research Progress in Biotechnology
Editor: G. E. Zaikov, pp. 53-58
ISBN: 978-1-60456-000-8

Chapter VII

Hypocholesterolic Effect of Some Higher Basidiomycetes

Alexey Popov*[1]*, Andrey Panchenko*[2]*, Nina Denisova*[1]*,
***Nikolay Petrishchev*[2] *and Mark Shamtsyan*[1*]**
[1] Department of Technology of Microbiological Synthesis, St.Petersburg State Institute of Technology (Technical University), 26 Moscovsky Pr., 198013 Saint Petersburg, Russia;
[2] Department of Pathological Physiology, Acad.I.P.Pavlov St.Petersburg Medicinal University, 6/8 L.Tolstoy Str.,197022 Saint Petersburg, Russia

Abstract

Submerge mycelia of various basidiomycetes were studied for hypocholesterolic action. Administration of mycelia of some of this mushroom accompanied with hyperlipidemic diet caused lowering of total cholesterol levels in blood and liver, increase of the level of α-cholesterol and decrease of the values of atherogenity coefficient and level of triglycerides in blood.

Keywords: Hyger basidiomycetes, submerge cultivation, hypocholesterolic effect.

Introduction

Mushrooms were attracting the interest of the people from ancient times. Edible mushrooms are attracting the interest of researchers first of all due to the prospects of obtaining of food protein and biologically active substances. Also the possibilities of submerge cultivation of higher basidiomycetes which are giving promising results, and which represents a significant interest for the biotechnology, are additionally increasing this interests /Chang, 1999, Hobbs, 1995, Wasser and Weis, 1999/.

Mushrooms possessing antisclerotic action, capable to decrease the level of cholesterine in blood, can be very perspective for medical industry and become a valuable diet product /, Wasser and Weis, 1999 , Gunde-Cimerman, 1999/.

During several recent decades, the problem of atherosclerosis and related diseases arises very strictly in the whole world.

The aim of our studies was definition of hypocholesterolic effect of food supplements from fruit bodies or submerge mycelia of several mushrooms.

Materials and Methods

The objects of our study weresubmerge cultivated higher basidiomycetes: Coriolus hirsutus, Cerrena unicolor, Trametes ochracea, Panus conchatus, Ganoderma lucidum, Flammulina velutipes, Coprinus logopides and Pleurotus ostreatus.

Submerge cultivation was performed on a culture media containing glucose as a carbon source and peptone as a source of nitrogen for 8 days. As a result of submerge cultivation necessary amount of biomass was collected, which was dried at 50°C, powdered and used for the studies of hypocholesterolic activity.

Determination of Hypocholesterolyc Activity of Basidiomycetes

The model of hyperlipidemia of rats has been chosen on the basis of experimental experience of Institute of experimental medicine of Russian Academy of Medical Sciences and other works. /Klimov and Rizhenkov, 1988, Staprans, et al., 1998/

Experiment was performed during 21 days using white male rats received from Rappolovo station of Russian Academy of Medical Sciences (St. Petersburg). Animals have been divided into groups on seven animals in each. Animals in intact group received a standard dry fodder and water *ad libitum*. The control group received hyperlipidemic diet and water. Animals from experimental groups received hyperlipidemic diet with the addition of mushroom preparation.

The composition of Hyperlipidemic diet was as follows: dry crystalline cholesterol -5%, pork fat -25%, sunflower seed oil -18%, vitamin D - 200 thousand units/day, the remaining – oat flakes. Hyperlipidemic fodder was prepared by mixing of the melted pork fat with crystalic cholesterol and with other above mentioned components. Animals received mushroom additives *per os*, mixed with a main bulk of a fodder in quantity of 2 % from a daily diet. Ammount of daily fodder for each animal was 20 grams. After 21 days the blood and a liver of rats was taken for analyses.

Determination of Total Cholesterol and Cholesterol of High Density Lypoproteids

For a rough quantitative estimation of a degree of risk of an atherosclerosis in 1977 Klimov A.N. offered so-called cholesteric factor of atherogenity (Kcl), representing a ratio of cholesterol of atherogenic lipoproteids (cholesterol of low density lipoproteids – Cl LDL and cholesterol of very low density lipopriteids – Cl VLDL) and cholesterol of anti-atherogenic lipoproteids (cholesterol of high density lipoproteids or an alpha cholesterol – Cl HDL) /Klimov and Nikulicheva, 1995/.

$$\mathrm{Kcl} = \frac{\mathrm{Cl\,LDL} + \mathrm{Cl\,VLDL}}{\mathrm{Cl\,HDL}}$$

As total quantity of cholesterol of atherogenic lipoproteids can be presented as a difference between the total cholesterol and cholesterol of high density lipoproteids, the factor of atherogenity can be calculated on the basis of definition only of two parameters - the total cholesterol and cholesterol of lipoproteids of high density:

$$Kcl = \frac{\mathrm{Cl\ total\ \ Cl\,HDL}}{\mathrm{Cl\,HDL}}$$

The concentration of total cholesterol and Cl HDL in blood plasma was defined using a cholesterol determination kit of "Human" (Germany). The method is based on a combination of enzymatic reaction with a reaction of oxidation. Peroxide of hydrogen formed as a result of these reactions due to the action of peroxidase reacts with 4-amynophenazon and phenol with formation of the colored product –chinonimin /Instruction of "Humana" cholesterol determination kit/.

Concentration of cholesterol was determined using following equation:

Ccl=14,3 x OD of sample (mol/l)

The concentration of total cholesterol in liver was determined using the following procedure: 200 mg of a liver tissue was dehydrated with 1,0 ml of methanol within 12 hours. Then added 2 ml of a mixture of heptan and isopropanol in a ratio 2:1. After 12 hours an extract was selected and evaporated. To a dry extract 0,1 ml of methanol and 2 ml of a standard reagent of "Human" company for definition of the total cholesterol were added. After 10 mines optical density density at 500 nanometers was measured. Concentration of cholesterol defined using above mentioned equation /Instruction of Humana cholesterol determination kit/.

Based on obtained data coefficient of atherogenity was culculated.

Results and Discussion

It is known, that hypercholesterolemia increases the risk of heart disease. Elevated levels of circulating cholesterol cause deposits to form inside blood vessels. These deposits can result in a disease process called *atherosclerosis*, which can cause blood clots to form that will ultimately totally stop blood flow. If this happens in the arteries supplying the heart, a heart attack will occur. If it happens in the brain, the result is a stroke where a portion of brain tissue dies. Atherosclerosis causes more deaths from heart disease than any other single condition. The most common cause of elevated serum cholesterol is eating foods that are rich in saturated fats or contain high levels of cholesterol.

Cholesterol has been divided into two major categories: low-density lipoprotein (LDL), the so-called "bad" cholesterol, and high-density lipoprotein (HDL), the so-called "good" cholesterol.

We studied the influence of mushroom additives to hypercholesterolic diets on the levels of total cholesterol and high-density lipoproteins in rats. The obtained results are demonstrating, that the addition of almost all of studied mushroom mycelia to hyperlipidemic diets of rats, caused significant decrease in a level of total cholesterol (fig. 5). Positive effect was not detected only for mycelia of Coprinus lagopides. In the cases of Trametes ochracea and Panus conchatus high levels of α-cholesterol in blood of experimental animals was almost normalizing the coefficient of atherogenity.

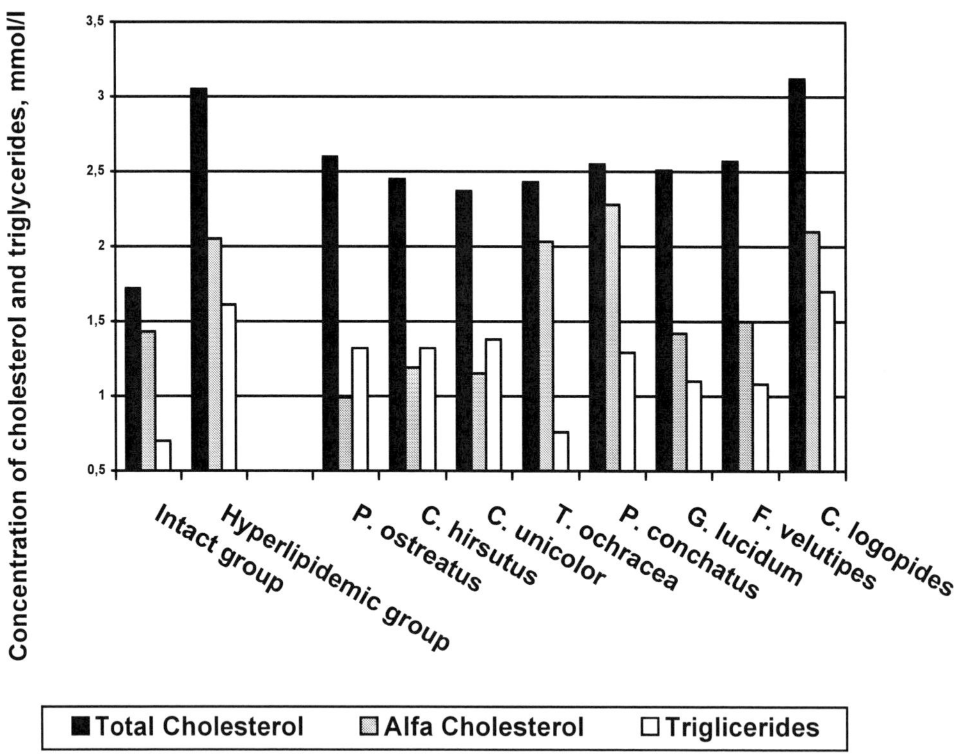

Figure 5. Concentations of Cholesterol and triglycerides in blood of different groups of rats.

Studies of the cholesterol levels in rat livers, shows that fruit mycelia of Cerrena unicolor, Pleurotus ostreatus, Trametes ochracea, Flammulina velutipes and Ganoderma lucidum were efficiently inhibiting the syntheses of cholesterol in liver (fig. 6).

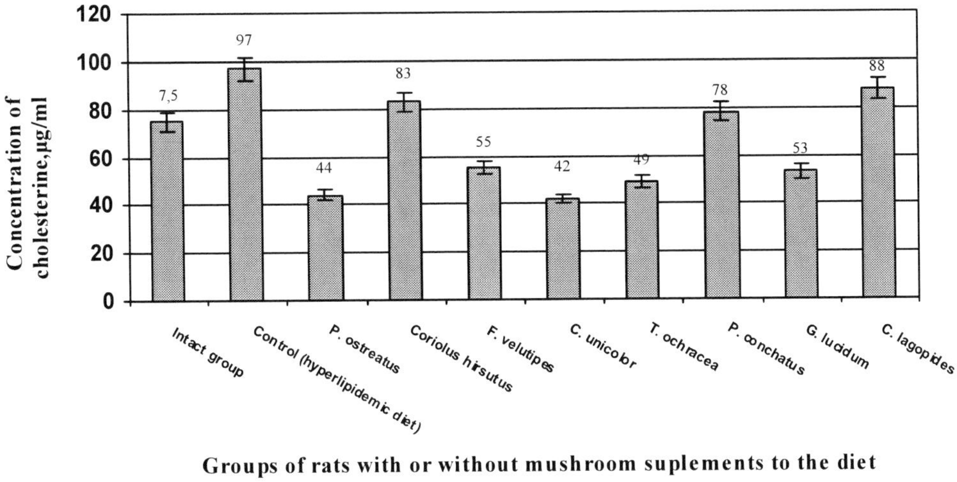

Figure 6. Contents of total cholesterine in the rat livers.

As a result of conducted studies it is possible to draw a conclusion that the additives to hyperlipidemic diets of a biomass of some mushrooms leads to essential decrease in a risk of development of atherosclerosis. The highest hypocholesterolic activity was detected for the fruit bodies of Pleurotus ostreatus and mycelium of Ganoderma lucidum.

Results of performed studies are demonstrating significant hypocholesterolic effect of uptake of supplements from various basidial mushrooms.

References

Chang, S.T. (1999). Global impact of edible and medicinal mushrooms on human welfare in the 21st century: nongreen revolution. *Int. J. Med. Mushrooms,* 1, 1-7.

Gunde-Cimerman, N. (1999). Medicinal value of the genus Pleurotus (Fr.)P. karst. (Agaricales S.I., Basidiomycetes). *International Journal of Medicinal Mushrooms* 1, 69-80.

Hobbs, C. (1995). Medicinal Mushrooms: An Exploration of Tradition, Healing and Culture. Botanica Press, Santa Cruz, CA. 347p.

Klimov A.N., Nikulcheva N.G. (1995). Lipids, *lipoproteids and atherosclerosis.* St. Petersburg, Chimia, 298p. (in Russian).

Klimov, A.N., Ryzhenkov, V.E. (1988). *Experimental studies of Hypolipidemic and anti-sclerotic preparations,* Moscow University, Moscow, 18 p. (in Russian).

Staprans, I, Pan, X-M, Rapp, J.H., Feingold. K.R., (1998). Oxidized cholesterol in the diet accelerates the development of aortic atherosclerosis in cholesterol-fed rabbits. *Arterioscler Thromb Vasc Biol.;*18, 977–983.

Wasser S.P., Weis A. (1999). Medicinal properties of substrates occuring in hygher Basidiomycetes mushrooms: current perspectives. *Int. J. Med. Mushrooms,* 1, 31-62.

In: Research Progress in Biotechnology
Editor: G. E. Zaikov, pp. 59-77
ISBN: 978-1-60456-000-8

Chapter VIII

Reaction of Ozone with Unsaturated Compounds: Concerted or Non-Concerted Addition

***B.E. Krisyuk*[*1] *and A.A. Popov*[*2]**
[1] Institute of Problems of Chemical Physics, Russian Academy of Sciences, Chernogolovka, Moscow District, Russia
[2] Emanuel Institute of Biochemical Physics, Russian Academy of Sciences, Moscow, Russia

1. Introduction

Interaction of ozone with a double bond is one of the most specific reactions of unsaturated compounds. This reaction is widely used in quantitative and qualitative analyses,

Olefin	$k(1)$ / share	$k(2)$	$k(1) + k(2)$	k_{exp}

synthesis, and chemical technology [1–3]. It is one of the best-studied reactions, which is described in hundreds of works and dozens of reviews and monographs published in the past 40 years [1–6]. Recently it is well known that ozone is a very important gas in the upper atmosphere. Also, the relevance of ozone for life process leads to importance in the experimental and theoretical studies of its reactions. The mechanism of this reaction was studied theoretically and experimentally during the last 50-60 years; until recently, it has been considered as an unambiguously established one. Many authors consider it as a classic example of simultaneous (concerted) 1.3-addition with the formation of five-membered cyclic molozonide or primary ozonide (PO) in the first stage (the Kriege mechanism [7]):

[*] B.E. Krisyuk: e-mail: bkris@mail.ru

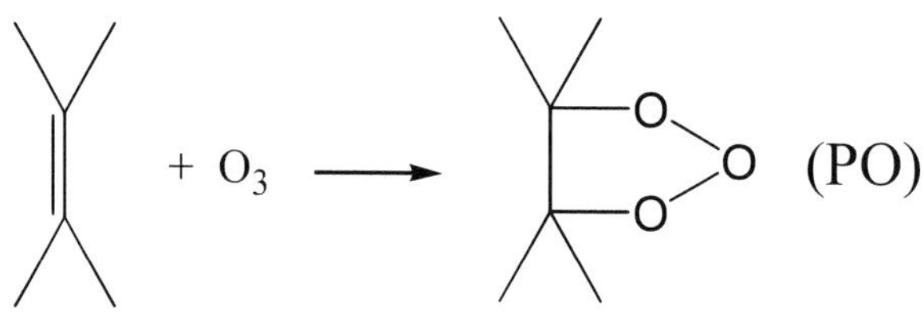

At moderate temperatures, PO is unstable and decomposes rapidly into a bipolar ion and a carbonyl compound (aldehyde or ketone), which, in turn, recombine and form "normal" ozonide:

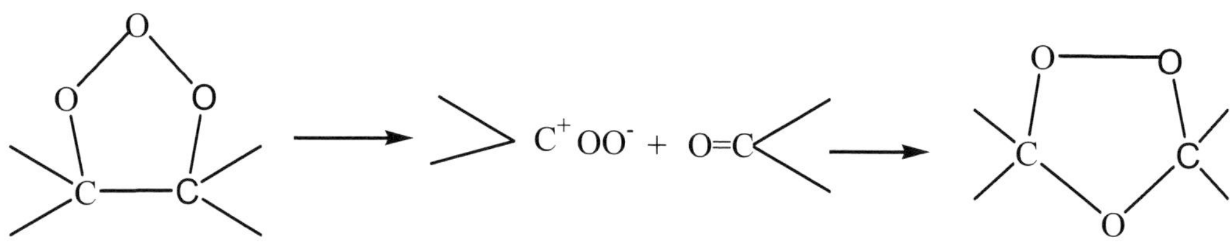

This reaction scheme was supported by numerous direct and indirect experimental data. At low temperatures, primary ozonide was obtained [8]; under the action of ozone on asymmetrical olefins, cross ozonides are formed [9]. This scheme and arguments for it were analyzed in detail in [1, 2]. But very limited kinetic information is available about atmospheric lifetimes of unsaturated compounds.

Some other schemes of the process were suggested, but they were rejected for various reasons. At present, the problem under discussion is whether the reaction proceeds as concerted addition or by two stages through the intermediate biradical. Most often, it is assumed that the former mechanism prevails for solutions; the latter one, for the gas phase [4–6, 10, 11]. However, at present, the mechanism of concerted addition is considered prevailing for the gas phase too [12–14].

The reaction mechanism, according to which ozone reacts with a double bond like a peroxyl radical and forms intermediate biradical, was suggested in [15] for the reaction with acetylene in an effort of explanation the difference in its Arrhenius parameters and parameters for the reaction of ozone with ethylene:

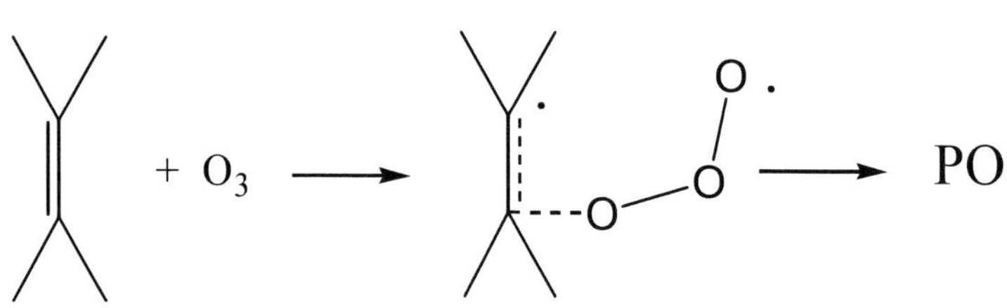

Later, the reaction with acetylene was studied by more advanced methods of quantum chemistry and this reaction mechanism was rejected [16]. However, this mechanism may explane the formation of such products as oxides and aldehydes wich usually considered as a

* A.A. Popov: e-mail: popov@sky.chph.ras.ru

result of decomposition of PO. In favor for this mechanism, other arguments may be suggested; an analysis of them was made in [17].

2. Modern Concepts on the Reaction Mechanism

At present, there is an opinion that the first stage of the reaction proceeds by another pathway. First of all, the reaction for ethylene and acetylene proceeds through formation of an intermediate weakly-bound complex, which was determined experimentally and theoretically [12–14, 16, 18]. Similar complexes were determined by methods of quantum chemistry for ozone with cis- and trans-butene-2 [19]. In [20], by approximation of the restricted Hartree–Fock method (RHF), the shape of the potential energy surface of the reaction of ozone with ethylene was calculated; it was shown that at the ground singlet state of the reagents in the region of transition state the reaction passing through asymmetric states is not energy-advantageous. In the concertedaddition, the potential energy surface of the reaction between the initial reagents and PO has two extrema: the minimum refers to the formation of a π-complex with an symmetrical structure and equal C...O (R_{CO}) distances of the order of 2.6–3.5 Å, and the maximum refers to the symmetrical transition state with R_{CO} of the order of 2 Å [12–14, 16–20]. The energy diagram of the reaction of ozone with multiple bonds is shown in figure 1. Thus, results of the recent studies are in favor of the mechanism of concerted addition with the formation of an intermediate complex of ozone with a multiple bond.

However, the calculations were performed with the restricted Hartree–Fock method (i.e., with a strictly-defined zero spin) in the one-determinant approximation. This approach may account for a loss of solutions for cases of complex systems where the spin squared deviates from zero although all reagents are in the singlet state. Hence, this solution may have the physical meaning [21], as in the case with ozone. In the literature, these cases are considered as unstable Hartree–Fock's solutions. In [22], calculations were performed with taking into account this condition in terms of the unrestricted Hartree–Fock (UHF) method; the calculations showed that the reaction may proceed by non-concerted addition but with the induction of higher-spin states (S^2 = 0.7–1.2).

In [23], another reaction pathway was analyzed: where ozone attacks ethylene with its central atom:

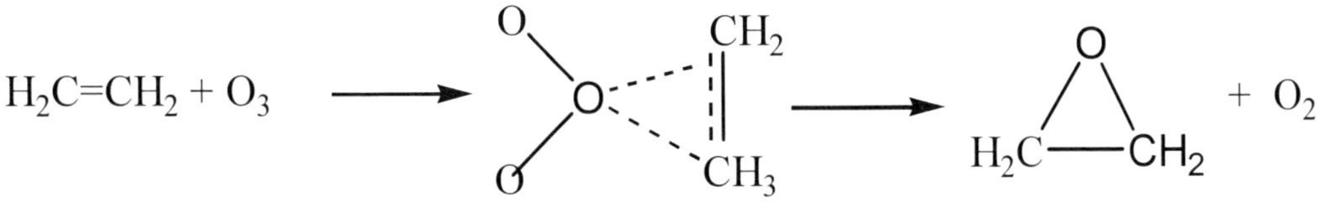

The authors [23] applied the density functional theory (exchange-correlation functional B3LYP) and calculated the profiles of the potential energy for the interaction of ozone with ethylene. It was shown that, according to the scheme of ozonolysis of ethylene, the reaction pathway through the direct epoxidation of the double C=C bond requires high activation energy and is improbable, both for thermochemical and photochemical reactions of ozonolysis.

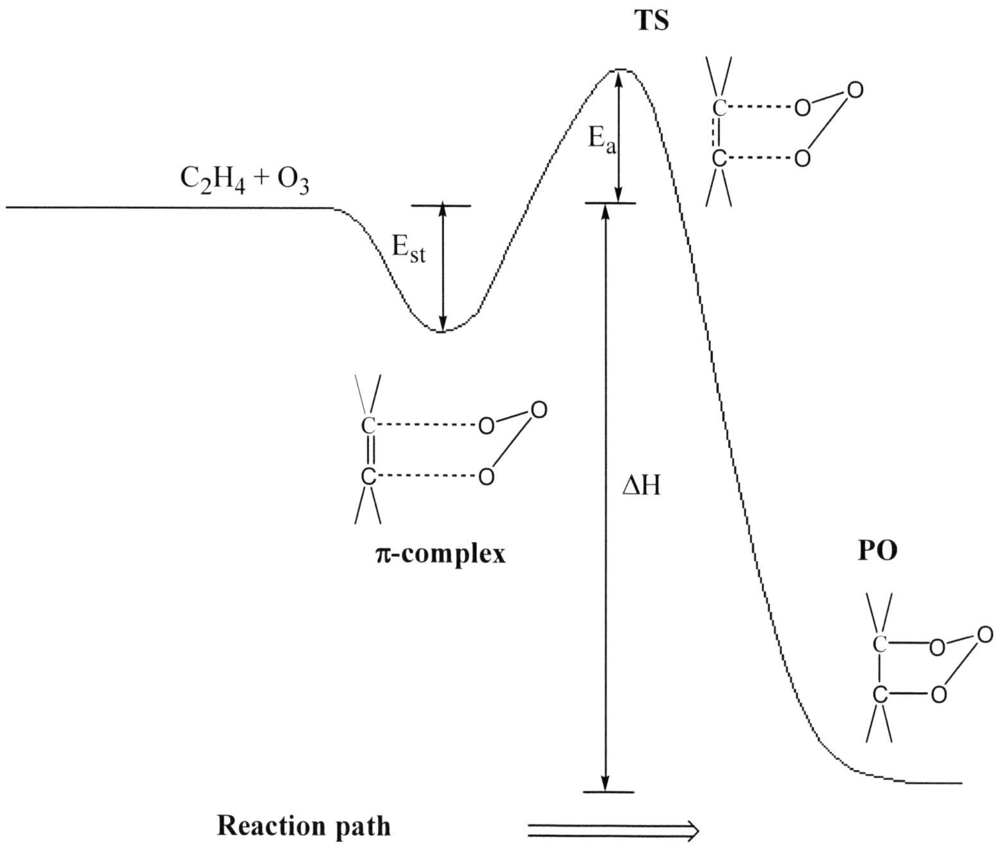

Figure 1. Energy diagram of ozone addition to ethylene.

In spite of the progress in investigating the mechanism of the reaction of ozone with multiple bonds, another pathway of the reaction mentioned in [16], in which ozone is the donor of atomic oxygen, has not been studied yet.

In many reactions, ozone behaves as a radical.[1] For example, the reaction of ozone with saturated hydrocarbons proceeds via H atom abstraction [25, 26]. This fact was supported by the following data [27]:

1. Ozonolysis of alkanes and polymers results in formation of free radicals.
2. Ozone initiates chain oxidation of hydrocarbons both in the liquid and gaseous phases.
3. The isotope effect correlates with the C-H and C-D bonds dissociation energies.

Consequently, ozone is an H atom acceptor similar to methyl radical or halogen atoms. This is evidence for the biradical nature of ozone and, hence, for a principal feasibility of the

[1] The ground state of ozone is a superposition of zwitterion and biradical; the share of the latter may reaches 100% [24].

reaction proceeding through an intermediate biradical state by analogy with peroxide radical reactions. It may be expected that ozone as biradical may readily tured into the triplet state; the mechanism of the reaction through triplet states has not been studied yet.

Thus, to sum up the above data and suppositions, the scheme of the first stage of the interaction of ozone with ethylene may be represented as follows:

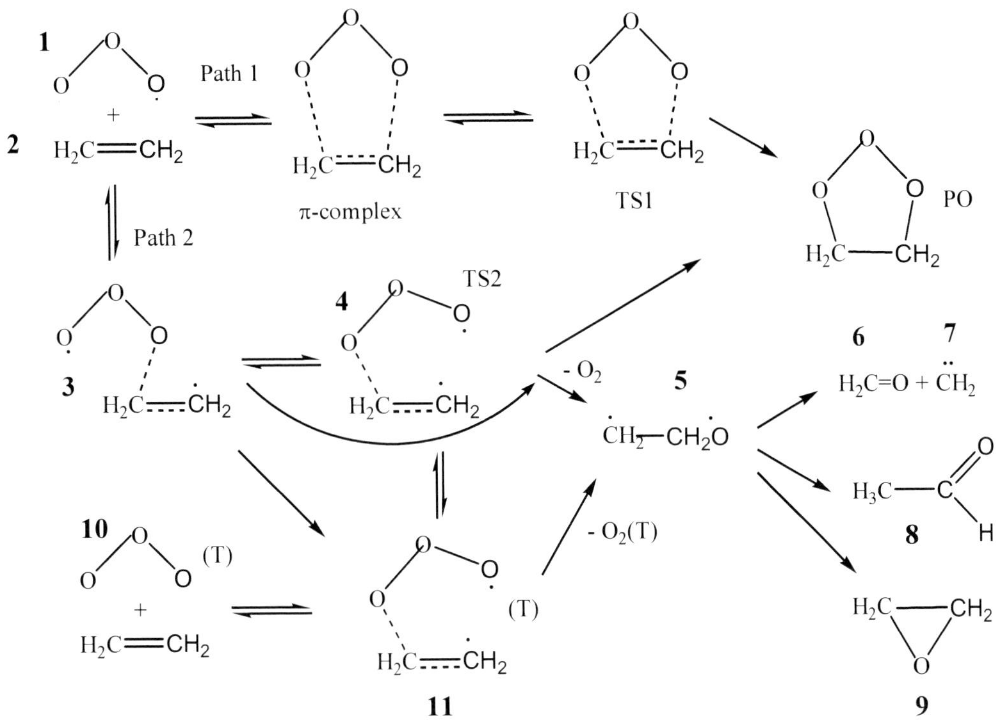

Here, path 1 (reaction 1) is the concerted addition of ozone (1) to ethylene (2), which proceeds through the formation of a weakly-bound complex that transforms into primary ozonide of ethylene (PO) or 1.2.3-trioxolene passing through the symmetrical transition state (TS1). Path 2 (reaction 2, the DeMore mechanism [15]) involves the collision during spontaneous orientation of the reagents (3) and the rotational transition to the biradical transition state (TS2) (4) followed by the formation of the same PO. Proceeding from the above-said, we supplement this pathway with the reaction of molecular oxygen elimination and the formation of intermediate biradical (5); the latter may either decompose with the formation of formaldehyde (6) and carbene (7) or transform into acetaldehyde (8) or epoxide (9). Finally, path 3 involves the transition of ozone into the triplet state (10). This pathway is similar to reaction 2. Here, the same biradical (5) is formed; it transforms into the products (6–8). The aim of this work is to compare different pathways of the reaction using data of the non-empirical quantum calculations.

3. Calculation Methods

The calculations were performed at the IPCP Computer Center on a CLI 16-processor cluster that incorporates four 4-processor units of 64-digit Intel Itanium 2 1.5 GHz processors with the cash 4 MB, 20 GB, HDD SCSI 3 x 146 GB 10,000 rpm. In the quantum calculations, we used a GAUSSIAN-03 program [28], *ab initio* HF and MP2 methods involving the restricted and unrestricted Hartree–Fock (RHF and UHF, respectively) methods, and the density functional B3LYP method involving the restricted and unrestricted Kohn–Sham methods (DFT analogs of RHF and UHF). In all cases, we applied a 6-31G basic sets family. The parameters of states corresponding to the minima at potential energy surface were determined by complete optimization of all variables; the transition states were determined from the complex and ozonide geometries. The zero point frequencies and energies were calculated at the extreme points of the potential plane. The calculations for the initial and transition states were performed by the MCSCF method. The thermodynamic parameters of the reaction were determined from results of the quantum calculations using the S.K. Ignatov's MOLTRAN program [29].

The shape of the potential energy surface in proximity to the transition state was determined by points. In this case, calculations were performed at given pairs of R_{CO} distances with other coordinates being optimized. The values of R_{CO} were varied from 1.8 to 3.1 A° for each pair of atoms (R_1, R_2); thus, a two-dimensional data file was obtained: the full energy as a function of two distances $U(R_1, R_2)$. Then, this data file was described by a surfase passing through the calculated points. To plot a graph, we passed a section through the plane obtained at a constant energy – the isoline.

4. Ozone+Ethylene Reaction

4.1. Initial Compounds

The ethylene molecule is stable with highly occupied orbitals. The triplet state is high and, as may be expected, the UHF calculations performed show the same result: the energy does not vary and remains in the singlet state with a zero spin. The total energy U and the reacting bond length R_{CC} calculated at different levels are presented in table 1.

For ozone, the situation is different. Here, solution of the one-determinant task yields two states: one with a zero S^2 and another with $S^2 = 0.4$–1.2 depending on the calculation method. Therefore, the UHF calculations yield different results. The UHF and UB3LYP calculations show a lower energy of the molecule, stabilization of solutions is observed. In this case, the states geometry is other than that determined by RHF and the spin squared is about 1 (UHF) or 0.4 (UB3LYP) (table 2). After the annihilation procedure, S^2 are 0.10 and 0.004, respectively. The MP2 method does not yield such solutions for ozone. The reason may be a known specificity of the MP2 method: it yields a too high energy of states with open shells.

Table 1. Parameters of ethylene

Method	Basis	R_{CC}, A	U, Hartree
HF	6-31G**	1,3165	-78,0388415
	6-31+G**	1,3208	-78,0430662
	6-311G**	1,3165	-78,0547235
	6-311+G**	1,3185	-78,0560772
B3LYP	6-31G**	1,3301	-78,5938076
	6-31+G**	1,3341	-78,5996455
	6-311G**	1,3269	-78,6139783
	6-311+G**	1,3288	-78,6155126
MP2	6-31G**	1,3352	-78,3172813
	6-31+G**	1,3392	-78,3231984
	6-311G**	1,3373	-78,3442916
	6-311+G**	1,3392	-78,3463029

Table 2. Parameters of ozone

Method	Basis	R_{OO}, A	α	S^2	U, Hartree
HF	6-31G**	1,204	119,0	0	-224,2614365
	6-31+G**	1,204	119,2	0	-224,2688212
	6-311G**	1,194	119,2	0	-224,3226419
	6-311+G**	1,194	119,4	0	-224,3296045
UHF	6-31G**	1,295	111,6	0,955	-224,3375691
	6-31+G**	1,293	111,8	0,956	-224,343609
	6-311G**	1,284	112,0	0,951	-224,3984805
	6-311+G**	1,284	112,2	0,953	-224,4052178
CAS	6-311G**	1,284	112,0	0	-224,4708544
triplet	6-311G**	1,241	131,2	2,064	-224,3093288
B3LYP	6-31G**	1,264	117,9	0	-225,4064536
	6-31+G**	1,263	118,1	0	-225,419503
	6-311G**	1,258	118,2	0	-225,4707934
	6-311+G**	1,256	118,4	0	-225,48056
UB3LYP	6-31G**	1,287	115,9	0,415	-225,4095947
	6-31+G**	1,285	116,3	0,394	-225,4222195
	6-311G**	1,281	116,3	0,408	-225,4738096
	6-311+G**	1,279	116,6	0,395	-225,483326
triplet	6-311G**	1,297	129,6	2,012	-225,4377512
MP2/UMP2	6-31G**	1,300	116,3	0	-224,8695447
	6-31+G**	1,301	116,5	0	-224,8861916
	6-311G**	1,282	116,8	0	-224,978195
	6-311+G**	1,282	117,1	0	-224,9916749
CAS	6-311G**	1,282	116,8	0	-224,9386337
triplet	6-311G**	1,274	129,3	2,076	-224,9042515

The ozone triplet state is higher than the stabilized singlet one at 80 kJ/mol (UB3LYP); its geometry differs from that of the singlet state. The valent angle increases to 130°. Here, $S^2 = 2$; as might be expected, after the annihilation procedure, S^2 remains unchanged.

The above results show that application of the one-determinant approximation to ozone in the *ab initio* methods results in significant errors and an abnormal spin squared. The DFT approximation is more efficient. The validity of the results may be verified using the multi-determinant approximation.

The results of MCSCF/6-311G** calculations for ozone (CAS(x,y) for "x" electrons and "y" MO) are shown below:

The MCSCF calculations showed that two MO in ozone are single-occupied, i.e., it is more advantageous for electrons to occupy spatially separated orbitals. This means that ozone in the basic state is singlet biradical. Hence, deviation of S^2 from zero is a result of solution of a two-centered task in one-determinant approximation; the solution is real and not a calculational artefact. Consequently, the UB3LYP results are reliable and we will analyze data obtained by this method.

Method	Energy, a.u.
RHF	-224,3226373
CAS(2,2)	-224,3297811
CAS(4,4)	-224,4323912
CAS(6,6)	-224,470855
CAS(8,8)	-224,5277802
MP2	-224,9612847

4.2. Complex

The approach of reagents before a transition state formation results in form of a weakly bound π-complex. According to the cited works, the complex structure is symmetrical with $R_1 = R_2 \sim 2.5 \div 3.6$ Å depending on the calculation method. The complex structure is shown in figure 2. However, an analysis of the complex showed that solutions relative to the RHF ↔ UHF transition are unstable both for the complex and initial ozone. The change-over to UB3LYP in the DFT calculations showed a lower total energy of the system and stabilization of solutions. The decrease in total energy of the symmetrical complex without optimization of its geometry is ~ 5 kJ/mol at the B3LYP level (table 3).

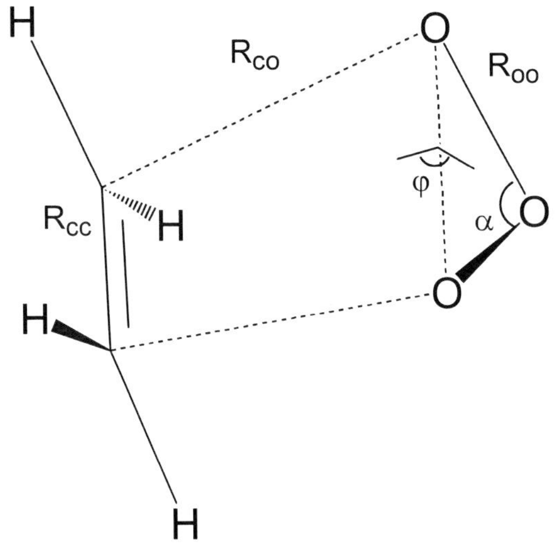

Figure 2. Structure of π-complex and transition state in the concerted (TS1) addition of ozone to ethylene.

Table 3. Parameters of ozone+ethylene complex

Method	Basis	R_1=R_2, A	R_{OO}, A	R_{CC}, A	α	S^2	*U*, Hartree
B3LYP	6-31G**	2,625	1,275	1,344	116	0	-304,00553
	6-31+G**	2,705	1,274	1,345	116,4	0	-304,022243
	6-311G**	2,66	1,267	1,339	116,6	0	-304,089716
	6-311+G**	2,75	1,265	1,339	116,9	0	-304,099168
UB3LYP	6-31G**	2,846	1,286	1,337	115,7	0,313	-304,006715
	6-31+G**	3,03	1,284	1,338	116	0,326	-304,023487
	6-311G**	2,874	1,279	1,333	116,1	0,313	-304,090941
	6-311+G**	3,041	1,278	1,333	116,4	0,332	-304,100568

4.3. Transition State

To verify whether the solutions for TS are unique or not and to determine the type of TS, we studied the shape of the potential energy surfase (PES) of the ozone–ethylene reaction. The shape of this surface in the TS region was determined by the HF [20], UHF, and UB3LYP methods [30] as a dependence of the energy *U* on two variables R_1 and R_2 (R_{CO} values for each pair of carbon and oxygen atoms).

In [20], the RHF calculations showed that the PES has one saddl point in the TS region; this point corresponds to symmetrical TS with $R_1 = R_2 = 2.165$ Å. The TS structure is similar to that of the π-complex; schematically, it is shown in figure 2 (TS1). Consequently, for the intreraction of ozone with ethylene considered in the one-determinant approximation (where $S^2 = 0$), the reaction pathway through the formation of symmetrical TS is more energy-advantageous.

Table 4 shows the energies and geometries of symmetrical transient states. Similar results were reported in [12–14, 20, 22].

Table 4. TS1 parameters

Method	Basis	$R_1 = R_2$	R_{OO}	R_{CC}	E_{act}, kJ/mol
HF	6-31G**	2,164	1,239	1,362	59,06
	6-31+G**	2,164	1,239	1,365	66,61
	6-311G**	2,149	1,232	1,365	63,31
	6-311+G**	2,153	1,232	1,366	68,69
CAS	6-311G**	2,149	1,232	1,365	269,85
B3LYP	6-31G**	2,348	1,288	1,359	-13,05
	6-31+G**	2,339	1,291	1,364	-6,56
	6-311G**	2,274	1,288	1,363	-10,43
	6-311+G**	2,276	1,289	1,366	-4,66
UB3LYP	6-31G**	2,348	1,288	1,359	-4,81
	6-31+G**	2,339	1,291	1,364	0,57
	6-311G**	2,274	1,288	1,363	-2,51
	6-311+G**	2,276	1,289	1,366	2,61
MP2(UMP2)	6-31G**	1,986	1,329	1,392	15,14
	6-31+G**	1,987	1,334	1,396	13,76
	6-311G**	1,965	1,311	1,397	27,29
	6-311+G**	1,973	1,313	1,399	24,88

The calculations with open shells showed a drastically different PES pattern. Here, the reaction pathway is different (through the asymmetrical TS with coordinates about 1.9 and 3.1); according to the UB3LYP data, PES has two saddles. Here, the reaction pathway from the complex to ozonide is feasible both through symmetrical and asymmetrical TS (figure 3).

Figure 4 shows the TS geometry determined by methods applicable to cases with open shells. This state is asymmetrical and of the biradical nature; this is evidence for the reaction proceeding by the mechanism described in [15]. This state may be referred to as TS2; its parameters are listed in table 5.

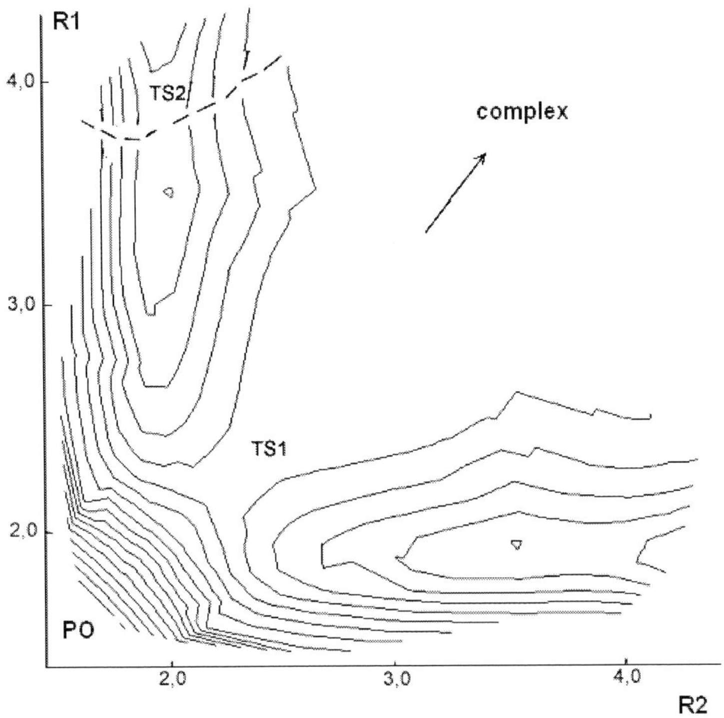

Figure 3. Potential enegy surface in the region of TS for the interaction of ozone with ethylene; UB3LYP calculations.

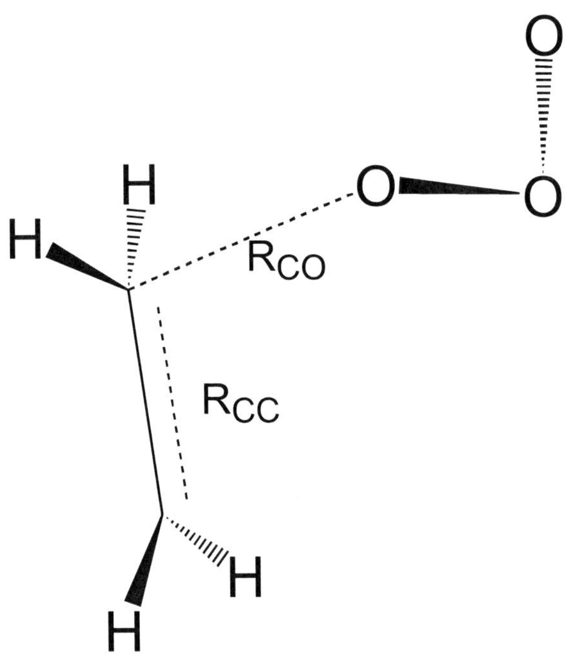

Figure 4. Transition state (TS2) in the non-concerted addition of ozone to ethylene.

Table 5. TS2 parameters

Method	Basis	R_1	R_2	R_{OO1}	R_{OO2}	R_{CC}	S^2	α	E_a, kJ/mol
UHF	6-31G**	1,888	3,125	1,350	1,28	1,398	1,221	110,7	67,62
	6-31+G**	1,893	3,177	1,348	1,279	1,399	1,217	111	73,13
	6-311G**	1,881	3,123	1,339	1,27	1,398	1,212	111,2	72,58
	6-311+G**	1,885	3,181	1,338	1,27	1,398	1,216	111,3	76,08
tripletr	6-311G**	1,913	3,922	1,350	1,28	1,394	2,26	109,7	35,96[2]
UMP2	6-31G**	2,019	3,051	1,369	1,287	1,342	1,207	112,1	207,18
	6-31+G**	2,024	3,150	1,370	1,286	1,344	1,202	112,5	212,57
	6-311G**	1,995	3,035	1,355	1,266	1,347	1,198	112,7	208,72
	6-311+G**	2,002	3,122	1,355	1,266	1,347	1,197	113	212,45
triplet	6-311G**	2,061	3,802	1,400	1,284	1,336	2.22	110,1	237,51[1]
UB3LYP	6-31G**	1,947	3,911	1,374	1,29	1,376	0,718	113,9	15,75
	6-31+G**	1,944	3,945	1,377	1,29	1,381	0,726	113,9	20,49
	6-311G**	1,926	3,902	1,379	1,282	1,377	0,740	114	19,74
	6-311+G**	1,926	3,902	1,379	1,282	1,377	0,740	114	23,49
triplet	6-311G**	1,926	3,902	1,379	1,282	1,377	2,02	114	52,22[1]

The calculations on the zero-point frequencies showed that both TS1 and TS2 have one imaginary frequency and are true transition states. Note that the TS2 geometry determined at all calculation levels is approximately the same but its energy determined by UMP2 and UB3LYP is higher (according to the HF data, lower) than that of TS1 (tables 4, 5). As is seen from tables 4 and 5, TS1 is a true singlet state ($S^2 = 0$) but TS2 has a higher spin squared ($S^2 = 0.7 \div 1.2$). After annihilating procedure, S^2 becomes nearly zero (0.005–0.03), like that for ozone. Figure 5 shows the energy profile of this reaction pathway and changes in S^2 associated with it. It is evident that S^2 increases gradually and reaches 1 at a distance of C...O about 1.55 Å; here, it intercrosses with the triplet surface.

The reaction on the triplet surface is similar in many respects to that through TS2 but here it is limited by the singlet–triplet transition. For the initial ozone, this transition requires considerable energy (about 80 kJ/mol) and it hardly probable for a thermal reaction. When the reaction proceeds in the light, the transition is feasible upon absorption of a quantum of visible light [1]. As the reaction proceeds, the ground and triplet states levels approach but the intercross of the surfaces occurs after TS2 (figure 5). According to the UB3LYP data, the reaction on the triplet surface proceeds without a barrier (figure 5) by elimination of triplet oxygen and formation of the same products as in the reaction with TS2 but without PO. It is the pathway, where ozone is a donor of molecular oxygen.

[2] Relative to singlet ozone

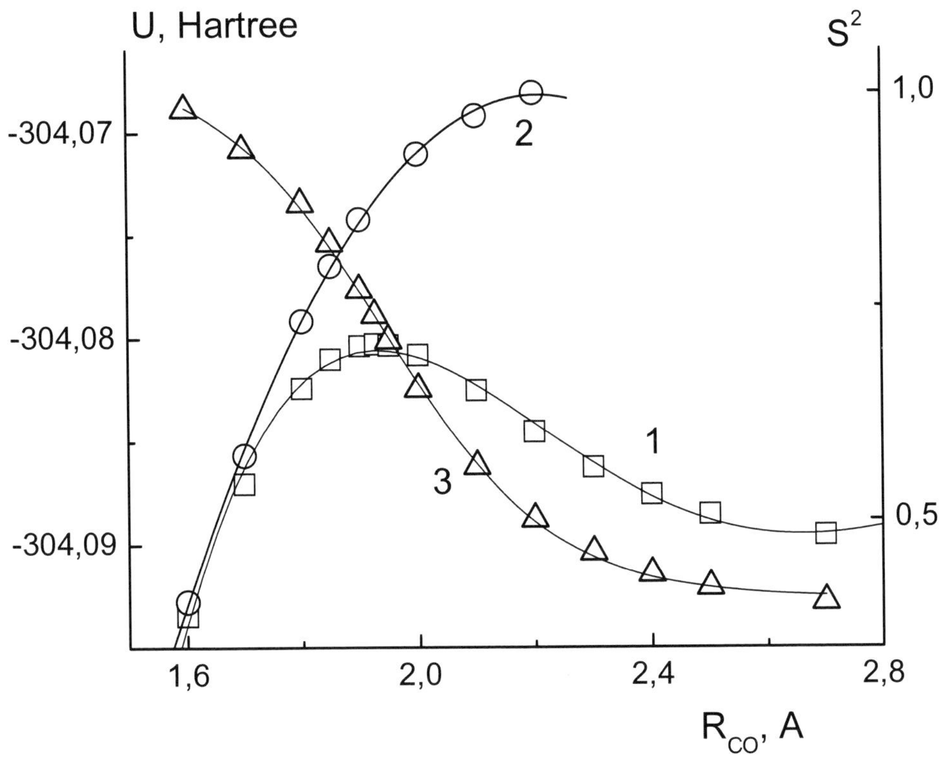

Figure 5. Change in the energy of the ethylene+ozone system by the pathway of non-concerted addition (1) in the singlet and (2) triplet states and (3) change in the S^2 value in the former case; UB3LYP calculations.

To verify the fact that the data obtained for TS2 are not an artefact, we performed the MCSCF calculations for both TS. In the calculations, we used from two to eight MO. The calculation results showed that the change-over to MCSCF results in a decrease in the total energy of reagents (ozone) and TS. (Tables 4 and 5 show the results obtained for six MO; the results for other variants are similar). Both for ozone and TS2, only two MO are single-occupied, which is evidence for the biradical nature of this TS. The anomalous value of S^2 obtained in the calculations performed in the one-determinant approximation is associated with this effect. For TS1, the MCSCF calculations showed that all MO are occupied by electron pairs; for this pathway, the one-determinant approximation is adequate. The MCSCF calculations showed that the energies of both TS are neighbour and both pathways may take place in parallel. Note that only the UB3LYP method describes adequately both TS.

The above data show that three pathways exist for the reaction of ozone with ethylene: concerted and non-concerted addition in the singlet state and a reaction in the triplet state. The latter pathway is hardly possible for the thermal reaction; the former two pathways exist in competition. To assess the efficiency of each pathway, we calculated the corresponding reaction rate constants. For this purpose, a MOLTRAN program and results of the quantum chemical calculations were used to calculate the enthalpies ($\Delta H^{\neq}$) and entropies ($\Delta S^{\neq}$) of activation of the two reaction pathways. From the data obtained, the corresponding rate constants k were calculated in terms of the standard transition state theory:

$$k = \frac{k_B T}{h} * \frac{RT}{P_0} * e^{-\frac{\Delta G^{*}}{RT}}$$

where P_0 is the standard pressure and h is the Planck's constant. The results obtained are listed in table 6.

As is seen from table 6, the reaction rate constants for both pathways are comparable and differ by 1 ÷ 2 order. This is evidence for simultaneous occurrence and competition of two pathways. The experimental reaction rate constant for 293 K is $1.77*10^{-18}$ cm^3/s*molec [31] or $1.07*10^3$ l/s*mol, i.e., approximately an average for these two pathways. On the whole, the assessment results point to the fact that the reaction for ethylene proceeds by 99% the mechanism of concerted addition, which is consistent with the conclusions of the above-cited works.

Table 6. Thermodynamic parameters of TS1 and TS2 ($\Delta H^{\neq}$ and $\Delta G^{\neq}$ in kJ*mol^{-1} , $\Delta S^{\neq}$ in J mol^{-1}deg^{-1}) and corresponding values of rate constants *k* (l^{-1}s^{-1}) at UB3LYP level

Reaction	Basis	$\Delta H^{\neq}$	$\Delta S^{\neq}$	$\Delta G^{\neq}$	k
(1)	6-31G**	5,60	-166,65	50,48	$2,15*10^5$
	6-31+G**	5,63	-166,71	55,90	$2,41*10^4$
	6-311G**	5,85	-170,24	54,09	$5,01*10^4$
	6-311+G**	5,72	-157,88	55,39	$2,96*10^4$
(2)	6-31G**	1,61	-149,58	61,95	$2,10*10^3$
	6-31+G**	1,53	-153,43	67,76	$2,01*10^2$
	6-311G**	1,62	-155,04	67,58	$2,16*10^2$
	6-311+G**	1,68	-135,66	65,62	$4,78*10^2$

5. Ozone + Tetrafluoroethylene Reaction

We studied the reaction of ozone with tetrafluoroethyne (TFE) by taking the suggested approaches. The reaction was analyzed in detail in [32], where it was shown that this reaction proceeds like that with ethylene. Here, the formation of a weakly bound complex also precedes TS. In this reaction, the PES structure in the TS region was similar to that for ethylene; but in the region of the complex it is almost plain. The surface has two saddles (two transient states) – TS1 and TS2. As is the case with ethylene, TS1 is symmetrical (R_1 = $R2$ = 2.3 Å); it corresponds to the reaction proceeding by the Kriege mechanism. The asymmetrical TS2 (R_1 = 2.0, $R2$ =4.0 Å) corresponds to the reaction proceeding by the DeMore mechanism.

Table 7 shows the electron contribution to the activation energy determined from a GAUSSIAN-03 and the Gibbs energy of transition state calculated with the aid of a MOLTRAN program from the zero oscillation frequencies. Also, table 7 shows the rate constant we determined for the symmetrical and asymmetrical reaction pathways and the experimental value borrowed from [33]. The latter value was obtained by the method of cryokinetic calorimetry (direct measurements on the rate constants of the reaction of TFE with pure ozone were performed at 90–150 K). The measurements showed that that the rate constant may be described as follows:

$$k \approx 2 \cdot 10^{-13} \cdot e^{-10000/RT}\ cm^3/s.$$

A comparison of the calculated rate constants for two reaction pathways showed that the reaction (2) constant determined for 150 K agrees better with the experiment; k for the reaction (1) is considerably lower. It should be noted that the k values showed in table 7 do not include possible rotation of the ozone fragment in TS2 (see the text above). Using a MOLTRAN program, we evaluated the contribution of this rotation to the TS entropy. With allowance for the rotation, $\Delta G^{\neq}$ changes to 1.3 kJ/mol, which results in 2- to 3-fold increasing the calculated values of k for the reaction with TS2; for the reaction with TS1, k does not vary. Hence, the asymmetrical reaction pathway is more advantageous.

The above data show that the reaction pathway (2) is more advantageous for the reaction of ozone with TFE. According to these data, the share of the reaction proceeding by the DeMore mechanism is not less than 99%.

Table 7. Energies of activation of the reaction E_a (at UB3LYP level), Gibbs energy of TS formation $\Delta G^{\neq}$ (kJ/mol), reaction rate constants k at 150K (l*(mol*s)$^{-1}$) for the reaction of ozone with TFE

Basis	E_a	$\Delta G^{\neq}$	k
Reaction 1			
6-31G**	5,9	58,08	$2{,}24 \cdot 10^{-7}$
6-31+G**	19,59	72,25	$2{,}59 \cdot 10^{-12}$
6-311G**	14,12	66,89	$1{,}90 \cdot 10^{-10}$
6-311+G**	21,95	74,79	$3{,}37 \cdot 10^{-13}$
cc-pvdz+	18,98	69,80	$1{,}84 \cdot 10^{-11}$
Reaction 2			
6-31G**	4,99	45,92	$3{,}84 \cdot 10^{-3}$
6-31+G**	14,19	56,90	$5{,}75 \cdot 10^{-7}$
6-311G**	9,07	51,83	$3{,}34 \cdot 10^{-5}$
6-311+G**	20,32	58,92	$1{,}14 \cdot 10^{-7}$
cc-pvdz+	15,99	56,93	$5{,}64 \cdot 10^{-7}$
Experiment[31]			$3{,}28 \cdot 10^{-7}$

6. Competition of Two Pathways for Reactions of other Compounds

In [17], two approaches were taken to evaluate the role of each reaction pathway: quantum-chemical calculations and parabolic model of the addition reaction (semiempirical method of intercrossing parabolas, MIP) [34-36]. Using these approaches in combination, the authors could evaluate independently the reaction rate constants for each pathway and compare their contributions to the total rate of ozonation of olefins of different structures. The comparison results are listed in table 8. We note a good agreement between the calculated and experimental constants values.

Table 8. Comparison of pathways 1 and 2 rate constants (l mol^{-1} s^{-1}, 300 K) calculated for reactions of olefins with ozone (liquid phase)

Olefin	$k(1)$ / share	$k(2)$	$k(1) + k(2)$	k_{exp}
CH_2=CH_2	1.17×10^5 / 100%	55.8	1.17×10^5	4.72×10^4
CH_2=CHMe	1.83×10^5 / 100%	17.8	1.83×10^5	1.40×10^4
CH_2=CHEt	1.83×10^5 / 100%	25.1	1.83×10^5	1.49×10^5
CH_2=CMe_2	6.45×10^5 / 100%	64.4	6.45×10^5	1.50×10^5
E-MeCH=CHMe	2.85×10^5 / 100%	32.1	2.85×10^5	
Olefin	k(1) / share	k(2)	k(1) + k(2)	kexp
Z-MeCH=CHMe	2.85×10^5 / 100%	96.7	2.85×10^5	1.89×10^5
MeCH=CMe_2	9.78×10^5 / 100%	1.67×10^2	9.78×10^5	5.80×10^5
Me_2CH=CMe_2	3.11×10^6 / 100%	2.11×10^3	3.11×10^6	1.58×10^7
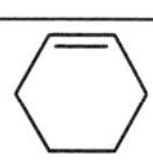	2.85×10^5 / 100%	5.43×10^3	2.86×10^5	8.90×10^5
CH_2=CHPh	3.99×10^3 / 32%	8.52×10^3	1.25×10^4	3.48×10^5
CH_2=CMePh	2.14×10^4 / 72%	8.14×10^3	2.95×10^4	
E-MeCH=CHPh	6.53×10^3 / 50%	6.50×10^3	1.30×10^4	
CH_2=CPh_2	1.55×10^3 / 2.7%	5.54×10^4	5.69×10^4	
E-PhCH=CHPh	2.65×10^2 / 0.02%	1.12×10^6	1.12×10^5	1.85×10^5
Z-PhCH=CHPh	2.65×10^2 / 0.00%	2.47×10^8	2.47×10^8	
CH_2=CHCH=CH_2	3.14×10^3 / 10%	2.80×10^4	3.11×10^4	8.01×10^4
E-CH_2=CHCH=CHMe	5.20×10^3 / 17%	2.60×10^4	3.12×10^4	

Table 8. (Continued)

CH_2=CMeCMe=CH_2	1.27×10^4 / 4.9%	2.46×10^5	2.59×10^5	3.25×10^5
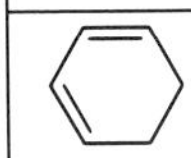	1.77×10^4 / 0.8%	2.18×10^6	2.20×10^6	

As it seen from table 8, from the point of view of the reactions (1) and (2) competition, unsaturated hydrocarbons may be subdivided into three groups.

(1) *Olefins.* Here, the reaction of symmetrical addition of ozone prevails. The reaction (2) share is less than 0.1%. The reaction (2) enthalpy for these olefins is -10 ÷ +6 kJ/mol^{-1}; thus, the reaction of symmetrical addition prevails. This conclusion agrees with the data of quantum-chemical calculations for ethylene and conclusions of many works on the reactions with olefins [1-3].

(2) *Styrenes and dienes.* The formation of π-bonds neighboring with the reacting site results evidently in increasing the classic potential barrier of the thermally neutral reaction; on the other hand, it makes the reaction (2) more exothermal (the enthalpy of such reactions vary over the range -35 ÷ -64 kJ/mol^{-1}). Hence, the reaction (2) makes a great contribution and, in cases of ozonation styrenes and dienes, it sometimes prevails.

(3) *Diphenylethylenes.* In reactions of ozone with diphenylethylenes, the reaction (2) prevails. The enthalpies of these reactions are low enough (they vary over the range -55÷-84 kJ/mol^{-1}).

7. Conclusion

Thus, the results presented show that the first stage of addition of ozone to multiple bonds may proceed by two different pathways on the singlet surface. The reaction on the triplet surface of potential energy is also probable. The ratio of the rates of the reactions (1) and (2) may vary over a wide range. The reaction (1) share in the overall ozonation is 100% for olefins. For TFE and diphenylethylenes the reaction (2) prevails. For styrenes and dienes, the reaction rates are comparable.

References

[1] Razumovskii S.D., Zaikov G.E. *Ozone and its reactions with organic compounds.* M., Nauka, 1974. (In Russian)

[2] Razumovskii S.D., Zaikov G.E. // *Uspekhi khimii,* 1980, vol 49, no., pp. 2344 – 2376. (In Russian)

[3] Lunin V.V., Popovich M.P., Tkachenko S.N. *Physical chemistry of ozone,* M., 1998. (In Russian)

[4] Bailey P.S. in: *Ozonation in Organic Chemistry.* vol. 1, 2. New York: Academic Press, 1968.

[5] Kuczkowski R.L. // *Acc. Chem. Res.* 1983. V. 16. N 1. P.42
[6] Bunnelle W.H. // *Chem. Rev.* 1991. V. 91. N . P.335.
[7] Criegee R. // *Angew. Chem.* 1975. V. 87. P. 765.
[8] Razumovskii S.D., Berezova L.V., *Izv. akad. nauk.* AN SSSR, Ser., 1968, p. 207. (In Russian)
[9] Murray R.W., Yossefueh R.D., Storey P.R. // *J. Amer. Chem. Soc.* 1967. V. 89. P. 2429.
[10] Cremer D. // *J. Amer. Chem. Soc.* 1981. V. 103. №. P. 3627.
[11] Ollzmann M., Kraka E., Cremer D., Gutbrod R., Andersson S.J. // *Phys. Chem.* 1998. V. 101. N . P. 9421.
[12] McKee M.L., Rohlfing C.M. // *J. Amer. Chem. Soc.* 1989. V. 111. № 7. P. 2497 – 2500.
[13] Gillies J.Z., Gillies C.W., Suenram R.D., Lovas F.J., Stahl W. // *J. Amer. Chem. Soc.* 1989. V. 111. № 8. P. 3073.
[14] Gillies C.W., Gillies J.Z., Suenram R.D., Lovas F.J., Kraka E., Cremer D. // *J. Amer. Chem. Soc.* 1991. V. 113. № 7. P. 2412 – 2421.
[15] DeMore W.B. // *Int. J. Chem. Kinetics.* 1969. V. 1.№ 1. P. 209.
[16] Cremer D., Kraka E., Crehuet R., Anglada J., Grafenstain J. // *Chem. Phys. Lett.* 2001. V. 347. P. 268 – 276.
[17] Denisov E.T., Krisyuk B.E., *Khim. fizika,* 2007, vol. 26. (In Russian) (In press)
[18] Gillies J.Z., Gillies C.W., Lovas F.J. ... // *J. Amer. Chem. Soc.* 1991. V. 113. № 17. P.6408-6415.
[19] Krisyuk B.E., *Zhurnal fiz. khimii,* 2004, vol. 78, no. 4, p. 1 – 5. (In Russian)
[20] Krisyuk B.E., Maiorov A.I., Popov A.a., *Khim. fizika,* 2003, vol. 22, no. 9, pp. 3-9. (In Russian)
[21] Zhidomirov G.M., Bagaturyants A.A., Abronin I.A., *Applied quantum chemistry, M., Khimiya,* 1979. 295 pp. (In Russian)
[22] Krisyuk B.E., *Khim. fizika,* 2006, vol. 25, no. 6, p. 13. (In Russian)
[23] Epoxidation
[24] Floriano W.B., Blazskowski S.R., Nascimento M.A.C. // *J. Molec. Struct.(Theochem.).* 1995. V. 335. P. 51.
[25] Krisyuk B.E., Polianchik E.V., *Khim. fizika,* 1990, vol. 9, vol. 1, p. 127. (In Russian)
[26] Timerghazin Q.K, Khursan S.L., Shereshovetz V.V. // *J. Molec. Struct.(Theochem.).* 1999. V. 489. P. 87.
[27] Denisov E.T., Denisova T.G. // *Russ. Chem.Bull.* 2002. .V 71. № 5. P. 417.
[28] Frisch M. J., Trucks G. W., Schlegel H. B., Scuseria G. E., Robb M. A., Cheeseman J. R., Montgomery J. A., Jr., Vreven T., Kudin K. N., Burant J. C., Millam J. M., Iyengar S. S., Tomasi J., Barone V., Mennucci B., Cossi M., Scalmani G., Rega N., Petersson G. A., Nakatsuji H., Hada M., Ehara M., Toyota K., Fukuda R., Hasegawa J., Ishida M., Nakajima T., Honda Y., Kitao O., Nakai H., Klene M., Li X., Knox J. E., Hratchian H. P., Cross J. B., Adamo C., Jaramillo J., Gomperts R., Stratmann R. E., Yazyev O., Austin A. J., Cammi R., Pomelli C., Ochterski J. W., Ayala P. Y., Morokuma K., Voth G. A., Salvador P., Dannenberg J. J., Zakrzewski V. G., Dapprich S., Daniels A. D., Strain M. C., Farkas O., Malick D. K., Rabuck A. D., Raghavachari K., Foresman J. B.,

Ortiz J. V., Cui Q., Baboul A. G., Clifford S., Cioslowski J., Stefanov B. B., Liu G., Liashenko A., Piskorz P., Komaromi I., Martin R. L., Fox D. J., Keith T., Al-Laham M. A., Peng C. Y., Nanayakkara A., Challacombe M., Gill P. M. W., Johnson B., Chen W., Wong M. W., Gonzalez C., and Pople J. A., Gaussian 03, Revision C.02, Gaussian, Inc., Wallingford CT, 2004.

[29] Ignatov S.K. Moltran v.2.5, Nizhny Novgorod, 2004, *http://ichem.unn.runnet.ru/tcg/Moltran.htm*

[30] Krisyuk B. e., Maiorov A.I., Popov A.A., *Khim. fizika,* (In press) (In Russian)

[31] Razumovskii S.D., *Khim. fizika,* 2000, vol. 19, no. 7, pp .58-62. (In Russian)

[32] Krisyuk B. e., Maiorov A.I., Popov A.A., *Khim. fizika,* (In press) (In Russian)

[33] Kiryukhin D.P., Barkalov I.M., Ismoilov I.L.,*Khim. fizika,* 2003, vol. 22, no. 2, p.123. (In Russian)

[34] Denisov E.T., Afanas`ev I.B. *Oxidation and Antioxidant in Organic Chemistry and Biology.* Taylor and Francis: Boca Raton, 2005. P. 101.

[35] Denisov E.T.,*Uspekhi khimii,* 2000, vol. 69, no. 2, p. 166. (In Russian)

[36] Denisov E. T., Models for Abstraction and Addition Reactions of Free Radicals in General Aspects of the Chemistry of Radicals, Z. B. Alfassi, (Ed.), John Wiley and Sons Ltd.: London, 1999, P. 79-137.

In: Research Progress in Biotechnology
Editor: G. E. Zaikov, pp. 79-85
ISBN: 978-1-60456-000-8

Chapter IX

Development of Technologies for Alternative Energy Sources Reception

M.G. Sagdieva*[*], *G.D. Zolotilina, G.V. Cherkasova, S.I. Borminskiy and H.M. Rajabov
Institute of Microbiology AS RUz, Tashkent, Republic of Uzbekistan
OAO "Almalyk GMK", Almalyk City, Republic of Uzbekistan

Abstract

From cattle manure in anaerobic conditions 31 association of methanogens - typical for methanic forms of microorganisms are isolated. In laboratory-scale plant with it's capacity 1,7 l technological parameters of processing of cattle manure to biogas are fulfilled with using isolated forms of methanic bacterium with cells titre of $6*10^9$ koE/ml at entering a sowing material of 5 % from volume: (a temperature optimum - 45-50^{0}C; crushing of organic substances, entry of nutrients to medium, maintenance of neutral pH value, mixing in anaerobic conditions).

On basis of received data pilot installation for getting biogas from cattle manure is created on the base of milk farm of Almalik mountain-metallurgical industrial complex (AGMK) with working volume 2 m^3. The steady yields of biogas about 1 m^3 of suspension 0,32-0,37 m^3 / day is obtained.

Keywords: Biogas, methanogens, anaerobiosis.

Literature Review

As a result of increase in fuel and energy source prices marked recently big attention is devoted to search of alternative energy sources, where biogas and bioethanol production have main interest. Methanogenesis is an original type of a metabolism that is carried out by

[*] E-mail: msagdieva@mail.ru

representatives of archaebacteria, allocating metane, as the main product of a metabolism formed at restoration of carbon dioxide.

In the literature the significant amount of works done for methanic forms of bacteria for transformation of manure to Biogas. Metane fermentation of cattle-breeding waste products has wide application in many industries and agriculture as an energy carrier since it has high fuel potential. However, process of metane formation by bacteria should pass in strict anaerobic conditions without access of oxygen, only in this case the most part of carbon is transformed to metane [1,2,3].

The purpose of our researches is isolation and selection of methanogens from cattle manure.

Materials and Methods

The quantitative calculation of activators in metane fermentation carried out on the medium from manure infusion with addition g/l: $(NH_4)_2SO_4$-0,5; peptone-1,0; $FeSO_4$-1,0; yeast extract -10, and on the synthetic medium of Pfenning [4] following mixture mg/l: NH_4Cl-330,0; KH_2PO_4-330,0; KCl-330,0; $MgCl*2H_2O$-330,0; $CaCl_2*6H_2O$-330,0; Na_2CO_3-1500; the microelements – 1ml, distilled water – 1l. A source of carbon - cellulose - 5.0 g either manure - 5,0 g or starch - 5,0 g. Amount of cells determined under the table of Mac Crady [5] after 7,17,21,28 days and expressed in koE/ml.

Isolation of accumulative methanogenic bacteria carried out in semi-fluid mediums (5 g of agar in 1l medium) at gas environment. Before crop, for absorption of oxygen reducer was added. Methanogens incubated at 40^0C within 30-90 days.

Preliminary clearing established on uniformity of colonies on dense and semi-fluid mediums.

Morphology of cells studied by using of microscopes MBI-3, MBI-6.

For detection of methane production laboratory installation with 1.7l capacity was designed.

Results of Researches

The analysis of microbiological studying of the quantitative contents of methanogens in cattle manure with overrotten straw laying and ground sediment from aryks with application of various mediums are submitted in table 1.

Table 1. The quantitative contents of methanogens in samples carbonaceous substrates on the medium from manure infusion and ground sediment

Nutrient medium	Cultivations								Amount of cells, koE/ml
	10^{-1}	10^{-2}	10^{-3}	10^{-4}	10^{-5}	10^{-6}	10^{-7}	10^{-8}	
cattle manure with straw laying	++	++	++	++	++	++	--	--	
ground sediment	++	++	++	+-	+-	--	--	--	

Table 2. The quantitative contents of methanogens in substrate of Pfenning (t=39^{0}C)

Variants of experience	Cattle manure	Ground sediments from aryk
Pfenning medium with a filtering paper, 5 g/l	2.5*10^5	2.5*10^4
Pfenning medium with starch, 5 g/l	2.5*10^4	2.5*10^4

As may be seen from data's submitted in table 1 and 2 the methanogens contents in samples with cattle manure + straw laying (2,5 million / ml) is detected.

Amount of methanogens on Pfenning medium with a filtering paper and starch on the order is lower.

Repeated passages of methanogens on organic and synthetic mediums have allowed to select 5 associations actively forming metane on manure. Results are submitted in table 3.

Intensity of growth of the selected associations, active methanogens is tracked depending on amount of a sowing material, temperature. The best results were given by association ZCH-3. It is established, that rise in temperature up to 50^{0}C stimulates growth of investigated bacteria and reduces terms of fermentation process. At entering of 5 % association ZCh-3 is revealed an intensification of a metagenesis, through 20-22 plentiful fermentation observed, which by 48 hours fades, and after 10 days it is marked again plentiful gas formation, number methanogens reaches 10^4-10^5 koE/ml, thus pH of medium is shifted in the alkaline side (pH 8,3-8,4), it is marked methan formation due to metabolites of fermentation by methanogens.

After 16-17 day cultivation the titre of methanogens has increased in 100 times and reached up to 10^6-10^7 koE/ml. The received data's are resulted in table 4.

In table 5 the data on change biogas yields from methane tanks are submitted at additional sowing fermented suspensions of cattle manure and mixes of cattle manure with straw by actively growing associations ZCh-3 and 17/4AG. In all variants brought associations developed on a background of natural microflora of manure.

Thus, on the basis of received data it is possible to make conclusions, that: from cattle manure associations of methanogenic bacteria are isolated, features of functioning of communities methanogens in system are investigated; the active community of bacteria is allocated, technological conditions for time reduction of gas – methane formation are picked up.

It is shown, that additional introduction to fermenting actively growing methanogene associations results in increase of speed and quantity of formed biogas.

Table 3. Associations methanogenic bacterias

Accumulative associations	A substrate from which the bacterium is allocated	Characteristics
17 AGM	Fermented manure sediment	Strains form flaky microcolonies which consist of bound cells. Cells under the microscope - thin slightly curved, sticks with roundish ends, found sticks with spores forming chains from 2-5 cells. Grow at 30-39^{0}C. Cells possess fluorescence in UV, and pH 7,0-8,9.
23 AGM	Fermented manure sediment	Strains form in solid and semi-fluid agar flaky microcolonies (starlets) and petty colonies with a dense center. At microscoping - cells as straight motionless sticks, single, or in pairs grow at 39-50^{0}C, pH 7,2-8,7 possess fluorescence in UV color.
ZCh-3	Fermented manure sediment+ overrotten straw	The association is submitted from three kinds of sticks: short, long, curved. In short sticks it is expressed sporogenesis, the association grows at 39-50^{0}C, pH 7,5-8,6, possess fluorescence in UV color.
17/4 AG	Fermented cellulose	The researched accumulative culture consist basically of petty sticks which are not painted on Gram, there are metasarcine microcolonies during active formation of gas are found. At cultivation in liquid mediums fine-dyspersated sediment is formed, from a sediment gas bubbles are allocated. The association grows in an interval 30-50^{0}C at pH7,0-8,9.
18/5 AG	Silt sediment of aryks	Units of cells and single coccus are observed. At microscoping isolate is submitted by packages from 2-4 cells, separate coccus and seldom found sticks. In the liquid medium grows as an whitish sediment. In agar mediums - granular colonies, grows in temperature 30-50^{0}C, pH 7,0-8,3.

Table 4. Influence methanogens associations on process of development Methanogenesis in laboratory installation

Entering of a sowing material, %	The first phase		The second phase		The third phase	
	Start of fermentation, days	Amount of methanogens, koE/мл	Plentiful gas formation, days	Amount of methanolgens, koE/мл	Plentiful gas formation, days	Amount of methanolgens, koE/мл
1	2-3 days after 3-4 days fermentation fades	10^2	14-15	10^3-10^4	23-24	10^5
5	20-22 hours, after 3-4 days fermentation fades	10^2	10	10^4-10^5	16-17	10^6-10^7
Control without adding methanogeneassociations	На 3-4 сутки	-	10-17	10^1	25-30	10^2

Table 5. The comparative data on efficiency of methane tanks (total yields of gases) at additional sowing before isolated and activated associations of microorganisms ZCh-3 (association I) and 17/4AG (association II) on a background of natural microflora development.

Added substrate	Used associations of microorganisms	Cultivation time, days											
		5		10		15		20		25		30	
		Gas yields		Gas yields		Gas yields		Gas yields		Gas yields		Gas yields	
		А	Б	А	Б	А	Б	А	Б	А	Б	А	Б
1	2	3	4	5	6	7	8	9	10	11	12	13	14
no	Natural microflora	13,38	0,110	6,21	0,050	21,39	0,171	20,08	0,161	30,37	0,243	35,63	0,285
	Additional sowing associations I	10,71	0,086	7,03	0,056	28,21	0,226	36,64	0,293	42,41	0,339	12,94	0,104
	Additional sowing associations II	5,95	0,048	4,72	0,038	15,46	0,124	38,51	0,308	40,59	0,325	8,6	0,069
Wheaten straw 10%	Natural microflora	11,29	0,090	5,23	0,042	23,42	0,187	41,16	0,329	25,08	0,201	9,33	0,075
	Additional sowing associations I	9,43	0,075	6,06	0,048	18,46	0,148	48,31	0,386	25,73	0,206	12,45	0,099
	Additional sowing associations II	12,03	0,096	5,98	0,048	13,59	0,109	44,06	0,352	39,78	0,318	8,53	0,068

A - amount of the water preemptibled at connection of methane tank to installation on metering of gas yields for three hours, in grammes (milliliters)

B - settlement productivity of methane tank l/l suspensions a day of m3 / m3 of suspension a day.

References

[1] Artamonov V.I. Biotechnology - to agriculture // *Moscow "Science",* 1989, p. 15;

[2] Vedenina I.Y., Slobodkin A.I. Characteristic of methane fermentation waste products of cattle-breeding complexes under the maintenance of a biomass of microorganisms and volatile fat acids // *Microbiology,* 1988, T.57, B. 4, pp. 695-697;

[3] Zhilina T.N., Zavarzin G.A. // *Microbiology,* 1979, pp. 451-456;

[4] Zhilina T.N., Ilarionova S.A. Characteristic formiate uptaking methane bacteria and description Methanobacterium thermoformicicum sp. Nov. // *Microbiology,* 1984. T.53, B.5, pp. 785-790;

[5] Methods of soil microbiology and biochemistry // under leadership Pof. Zvjagintseva D.G., Moscow State University publishing house, 1991, p. 302;

In: Research Progress in Biotechnology
Editor: G. E. Zaikov, pp. 87-97
ISBN: 978-1-60456-000-8

Chapter X

Polyfunctional SAS of Microbial Origin for Agriculture

M.G. Sagdieva, G.V. Cherkasova, G.I. Djumaniyazova, A.E. Kurbanbaeva, S.D. Gusakova and Zh.S. Ayropetova
Institute of Microbiology, Uzbekistan Academy of Sciences,
Uzbekistan, Tashkent, 100128, A. Kadyriy Street, b. 7[b]

Abstract

On the basis of screening 19 strains of nitrogen-fixing and 9 strains of phosphorous-mobilizing bacteria, isolated from rhizosphere of cotton, alfalfa and sugar-beet, two active strains - producers of biosurfactants (BS) - have been isolated and identified as *Pseudomonas aeruginosa CH5* and *Bacillus subtilis 26*. The value of surface tension of cultural broth (CB) of the mentioned bacteria varied within the limits 57.2 – 58.0 mN/m. The optimal cultivation conditions for active production of BS by *P. aeruginosa CH5* and *B. subtilis 26* have been determined, at which the surface tension of CB had value 36.6 – 40.5 mN/m. From CB of *P. aeruginosa CH5* biomass, grown in BEB (beef-extract broth) with addition of sunflower or cotton oil, the BS complex (0.2-2.1%) has been selected. In composition of BS no less than three rhamnolipids and lipophilic pigments of phenazine nature have been found, the amount of which essentially depended on type of vegetative oil in nutrient medium. CB, BS and water-soluble pigments of *P. aeruginosa CH5* displayed growth-stimulant activity towards seeds and seedlings of above-mentioned agricultural crops and antagonistic activity to *Verticillium dahliae, Rhizoctonia solani* and *Fusarium oxysporum* phytopathogenic fungi. This activity remained after BS autoclaving (121°C, 1 atm.), enduring at pH 5.5-8.5 and NaCl concentration up to 20 g/L. CB, BS and water-soluble pigments were stable during storage in usual conditions more than 9 months.

Keywords: *Pseudomonas aeruginosa*, *Bacillus subtilis,* Rhizobacteria, Biosurfactants, Rhamnolipids, Phenazine pigments, Bioremediation.

Introduction

At present time in connection with considerable contamination of land resources and loss of their aptitude for irrigated agriculture it is necessary to conduct the search of microorganisms, which are able to survive in stress conditions, improve soil properties and fertility and simultaneously to influence on growth and development of plants [1]. The bioremediation with application of microorganisms, producing microbial SAS (biosurfactants, BS) and their enzymes, which are active towards organic and inorganic pollutants, is a modern method to restore poorer and contaminated soils. Microbiological method of clean-up environment and soil restoration should organically combine with natural balance in ecosystems. The main demand of acceptability of soil bioremediation technologies is a complete preservation of population health, inhabiting this area, and also restoration of favorable ecological conditions [2-4].

The advantage of microbial BS over synthetic ones includes their non-toxicity, ecological compatibility, easy utilization, stability to extreme temperatures, overpressure, relatively high values of pH and salinity, possibility of their obtaining by biotechnological way. The microbial BS display immunomodulating and antagonistic properties against phytopatogens [5]. However, the investigations devoted to associative interaction of exometabolites from rhizobacteria including BS with high plants are single.

One of factors limiting wide application of microbial BS is relatively low number of microorganisms-destructors in soils, whereas in rhizosphere zone their number is ten times more. Rhizosphere soil differs on quality and quantity from microflora of control. Since rhizosphere microorganisms excrete bioactive substances into the soil including BS, it is of great interest to use them for soil bioremediation, creation of biological preparations of polyfunctional action, which are able to stimulate plant growth and protect them from phytopathogens [2].

It is known that species of *Pseudomonas* genus *(P. aeruginosa, P. chlororaphis, P. putida)* produce active BS of glycolipid nature – rhamnolipids, amount of which depends on strain and source of carbon [6]. So, cultivation of *P. aeruginosa* in media, containing additions of vegetative oils as source of carbon (corn oil, olive oil, soybean or canola oil), promotes of accumulation of rhamnolipids in CB from 7.65 to 112 g/L [7]. Such exametabolites of bacteria as pigments can be extracted from biomass together with BS complex. Different strains of *P. aeruginosa* synthesize two main blue phenazine pigments – pyocyanin (1-hydroxy-5-N-methyl-phenazine) and phenazine-1-carboxylic acid. Pyocyanin belongs to high-active antibiotics, it has toxicity and practical application, because it is active against pathogenic bacteria. The close correlation between secondary metabolites, synthesized with bacteria, has been recorded, in particular, if pigments are synthesized, then production of both antibiotics and other bioactive substances occur [8].

The objective of investigations was a search of local industrial-perspective strains of nitrogen-fixing and phosphorous-mobilizing bacteria – producers of BS, as well as selection of optimal conditions of cultivation and characteristic of BS.

Materials and Methods

Microorganism

Isolation of rhizobacteria – producers of BS was conducted on general accepted methods of soil microbiology from rhizosphere of wheat, cotton, alfalfa and sugar-beet. The species identification of microorganisms was determined on Bergey's Manual of Determi-native Bacteriology [9]. Active strains were selected on the basis of analyzing results of CB surface tension value measurement by methods of Rebinder and Wilhelmy plate technique [10]. The active strains were isolated in pure culture in a way of successive passages to agarized BEB (beef-extract broth) at temperature 32-37°C.

Conditions of Cultivation

The BEB, BEA and Chapek`s media were used. According to the value of surface tension of CB, the optimal nutritive medium (BEB + 1% cotton or sunflower oil) was selected. The conditions of cultivation in this medium were as follows – (orbital shaker, 180 rpm, 37°C) and age of culture (7 days and more) – when the surface tension of CB had the lowest value. Phytopathogenic fungi were cultivated in Chapek`s medium.

Determination of Activity

The influence of both CB and BS, isolated from active cultures, was studied in different concentrations on the seeds of above-mentioned cultures in Petri dishes with filter paper. To obtain this, the seeds were sterilized with concentrated H_2SO_4 and introduced into Petri dishes on 50 pieces into each dish, followed with introduction on 10 ml of CB in concentration 10^6cells/ml or BS in concentration 0,01- 0,00001%. The seeds, impregnated with 10 ml of sterile tap water, were served as control. The germination and rate of sprouting seeds were registered during 4 - 6 days.

Antagonistic activity of bacteria to phytopathogenic fungi *Verticillium dahliae, Rhizoctonia solani* and *Fusarium oxysporum* was determined with method of holes in agar. 7-daily suspension of phytopatogenic fungi was sowed as lawn to the surface of agarized Chapek`s media. 7- or 15-daily CB (10^9 cells/ml) or 1% emulsion of BS were introduced into each hole cut out in the center of agar plate (d = 0.7 cm). The incubation was conducted at 28°C.

Rhamnolipids Extraction and Characterization

250 ml of CB of *P. aeruginosa CH5* (pH 7.0, σ = 36.6-40.4 mN/m) were centrifuged at 7,000 g for 30 min. In the first experiments the supernatant was acidified with 10% HCl up to pH 2.0-3.0, it was undergone +5°C during night and further it was extracted 3 times on 70 ml

of mixture of chloroform and methanol (2:1, v/v) [11]. The united extracts were dried up over dry Na_2SO_4 and steamed on rotary evaporator. Because of absence of well-formed sediment of rhamnolipids within acid solution the next samples of supernatant were extracted without acidification.

The complex of BS in the form of brown pasty substance was analyzed with thin-layer chromatography (TLC) method on glass plates (9 x 12 cm) with silica gel of grade L5/40 and addition of 13% Ca_2SO_4 (Cavalier, Czechoslovakia) in system of solvents $CHCl_3$-MeOH-AcOH (65:15:2, v/v/v). The spots were developed with iodine fumes, and also spraying of chromatogram with 0.5% solution of α-naphthol in H_2O-MeOH (1:1, v|v), then after spraying with 50% H_2SO_4 and heating up to appearance of violet color of glycolipids spots [12]. The spots of rhamnolipids were identified on the basis of comparison of R_f with literature data [13,14]. The quantitative estimate of the main types of rhamnolipids of 15-daily culture of *P. aeruginosa CH5*, grown up on sunflower oil, was conducted on data of double TLC under mentioned conditions, for 7-daily culture grown up on cotton oil – by method column chromatography (CC) [11]. After removal of lipophilic BS from supernatant its surface tension was measured, then 50 ml was sampled, dried until dry and water-soluble BS content was determined gravimetrically.

Results and Discussion

Since in rhizosphere soil there are microorganisms by order higher and more diverse types than in control soil [2], i.e. the search of perspective microorganisms – producers of BS was conducted in rhizosphere zone of cotton, alfalfa, wheat and sugar-beet. 19 strains of nitrogen-fixing and 9 strains of phosphorous-mobilizing bacteria have been revealed, from which two strains from wheat rhizosphere have been found as active producers of BS. They were identified as *Pseudomonas aeruginosa CH5* and *Bacillus subtilis 26*. The CB of the mentioned bacteria had value of surface tension within limits 57.2 – 58.0 mN/m. *P. aeruginosa CH5* strain was cultivated in choosed optimal conditions in the medium with BEB + 1% cotton or sunflower oil during 7-15 days. The obtained CB was analyzed on content and composition of BS.

General Characteristics of BS Complex from *P. Aeruginosa* CH5

Growing *Pseudomonas* genus strains in the nutritive media with addition of vegetative oils promotes to synthesis of rhamnolipids [7]. For cultivation of *P. aeruginosa CH5* strain we used cotton and sunflower oils as carbon source. The latter was chosen as the most economical for industrial cultivation of strain. The accumulation of BS in CB was estimated in dynamics of culture growth on value of surface tension. The lowest index was found in CB of 7-daily and 15-daily culture, grown up on sunflower oil, and in 7-daily culture grown on cotton oil. During storage of CB, obtained on cotton oil, during 20 days no noticeable changes of σ were observed. From supernatant of these samples the lipophilic BS (LBS) and water-soluble BS (WBS) were isolated. The results of experiments are represented in table 1, from data of which it is seen that under equal age of cultures (7 days) and comparable σ values of CB and supernatant, the LBS and WBS content is higher than in supernatant from cotton oil. The removal of LBS from supernatants does not increase essentially their surface tension. Content of exometabolites (LBS+WBS) in CB increases with growing age of culture (7-20 days), mainly at the expense of WBS.

BS of *P. Aeruginosa CH5* Culture, Grown in Sunflower Oil

CB and supernatant of these samples had brown colour. According to results of analytical TLC, LBS of 15-daily culture were a mixture of no less than 5 types of rhamnolipids with pigments. Spots of pigments were coloured in yellow, green and brown colors.

On the basis of specific violet color of glycolipids spots after development of chromatogram with α-naphthol and comparison of R_f values with literature data [13,14], rhamnolipid RL-4 (R_f 0.16; 2 molecules of rhamnose, 1 molecule of fatty acid 3-OH-10:0) was identified as main component, rhamnolipids RL-3 (0.33; 1 molecule of rhamnose, 1 molecule of fatty acid 3-OH-10:0) and RL-2 (0.41; 2 molecules of rhamnose, 2 molecules of fatty acid 3-OH-10:0) – as minor components. Other minor components with R_f 0.07 and 0.27 were not identified. Amount of rhamnolipids in complex with LBS (data of preparative TLC) comprised 49%, from them RL-4 – 38.6; RL-3 – 6.6; RL-2 – 3.8% .

Thus, RL-4 and RL-3 are the main rhamnolipids of 15-daily culture of *P. aeruginosa CH5*. In literature the structures of 4 types of rhamnolipids with residuum of one or two molecules of L-rhamnose and one or two molecules of 3-hydroxydecanoic fatty acid (3-OH-10:0) are described. Two types are mains, and two types are accumulated in dependence on cultivation conditions for microorganisms [7, 13]. It has been noted that both RL-4 and RL-3 rhamnolipids are synthesized biosynthetically only under some conditions of cultivation [13].

BS of *P. Aeruginosa CH5* Culture, Grown in Cotton Oil

CB and supernatant of these samples had dark-brown color. The composition of rhamnolipids of 7-daily and 20-daily culture in comparison to analogous one of he culture, grown in sunflower oil, was more simple (analytical TLC) and included mainly both RL-4 and RL-3 with prevalence of the latter one. The set of lipophilic pigments (R_f 0.7-0.9) differed with complexity and color diversity (pink, blue, green, different tints of brown color). According data CC, content of RL-4 and RL-3 rhamnolipids in 7-daily CB comprised 4.8% and 6.8% correspondently, but pigments content – more 85%. It is known that the main blue phenazine pigment of *P. aeruginosa* – pyocyanin – can exist in oxidated and reduced forms. The reduced form is unstable free radical and it reacts quickly with molecular oxygen with formation of coloured compounds [15, 16]. The accumulation of phenazine pigments in *P. aeruginosa CH5* culture, grown in cotton oil, can be explained both specificity of strain itself and composition of lipophilic components of cotton oil, for example gossypol.

Table 1. Characteristic of BS complex from *P. aeruginosa CH5* under growing of culture in the media with addition of vegetative oils

Age of culture	σ, mN/m		Output of LBS from supernatant		σ of super-natant after extraction, mN/m	Output of WBS from supernatant		Total output of LBS and WBS	
	CB	Supernatant	g/L	%		g/L	%	g/L	%
Sunflower oil									
7-daily	36,60	36,80	0,67	0,07	40,02	8,1	0,81	8,77	0,88
15-daily	36,90	-	2,14	0,21	-	-	-	-	-
Cotton oil									
7- daily	37,19	38,06	0,44	0,04	37,58	6,2	0,63	6,64	0,66
20-daily	40,50	40,10	0,42	0,04	40,45	15,5	1,55	15,92	1,59

BS of *P. Aeruginosa CH5* Culture, Grown in Sunflower Oil

CB and supernatant of these samples had brown colour. According to results of analytical TLC, LBS of 15-daily culture were a mixture of no less than 5 types of rhamnolipids with pigments. Spots of pigments were coloured in yellow, green and brown colors.

On the basis of specific violet color of glycolipids spots after development of chromatogram with α-naphthol and comparison of R_f values with literature data [13,14], rhamnolipid RL-4 (R_f 0.16; 2 molecules of rhamnose, 1 molecule of fatty acid 3-OH-10:0) was identified as main component, rhamnolipids RL-3 (0.33; 1 molecule of rhamnose, 1 molecule of fatty acid 3-OH-10:0) and RL-2 (0.41; 2 molecules of rhamnose, 2 molecules of fatty acid 3-OH-10:0) – as minor components. Other minor components with R_f 0.07 and 0.27 were not identified. Amount of rhamnolipids in complex with LBS (data of preparative TLC) comprised 49%, from them RL-4 – 38.6; RL-3 – 6.6; RL-2 – 3.8% .

Thus, RL-4 and RL-3 are the main rhamnolipids of 15-daily culture of *P. aeruginosa CH5*. In literature the structures of 4 types of rhamnolipids with residuum of one or two molecules of L-rhamnose and one or two molecules of 3-hydroxydecanoic fatty acid (3-OH-10:0) are described. Two types are mains, and two types are accumulated in dependence on cultivation conditions for microorganisms [7, 13]. It has been noted that both RL-4 and RL-3 rhamnolipids are synthesized biosynthetically only under some conditions of cultivation [13].

BS of *P. Aeruginosa CH5* Culture, Grown in Cotton Oil

CB and supernatant of these samples had dark-brown color. The composition of rhamnolipids of 7-daily and 20-daily culture in comparison to analogous one of he culture, grown in sunflower oil, was more simple (analytical TLC) and included mainly both RL-4 and RL-3 with prevalence of the latter one. The set of lipophilic pigments (R_f 0.7-0.9) differed with complexity and color diversity (pink, blue, green, different tints of brown color). According data CC, content of RL-4 and RL-3 rhamnolipids in 7-daily CB comprised 4.8% and 6.8% correspondently, but pigments content – more 85%. It is known that the main blue phenazine pigment of *P. aeruginosa* – pyocyanin – can exist in oxidated and reduced forms. The reduced form is unstable free radical and it reacts quickly with molecular oxygen with formation of coloured compounds [15, 16]. The accumulation of phenazine pigments in *P. aeruginosa CH5* culture, grown in cotton oil, can be explained both specificity of strain itself and composition of lipophilic components of cotton oil, for example gossypol.

Biological Activity of CB and BS from *P. Aeruginosa CH5*

The action of CB and BS in concentrations from 0.01 to 0.00001% on germination and sprouting seeds as well as to antagonistic activity to phytopathogenic fungi was studied. Comparison of activity before and after autoclavating showed that autoclaving did not influence on properties of CB and BS, because their activity remained. Data of these experiments are shown in tables 2 and 3.

Table 2. Seed germination of cotton and wheat, treated with 15-daily CB of *P. aeruginosa CH5*

Sample	Time of incubation, (days)	Control-1 (H_2O), %	Control-2 (BEB of 6th dilution), %	CB 10^6 cells/ml, %
Cotton				
Seeds	1	10	52	30
	2	50	80	66
	3	72	85	86
	4	88	85	94
Seedlings (length, cm)		2,14	2,0-3,0	4,81
Wheat				
Seeds	1	56	40	82
	2	76	72	90
	3	78	75	96
	4	82	75	100
Seedlings (length, cm)		0,5-5	0,5 - 5	9,0 – 9,5

Table 3. Seed germination of cotton and wheat, treated by initial and autoclaved BS complex of 15-daily *P. aeruginosa CH5* culture

Sample	Time of incubation, days	BS concentration before autoclaving, %		BS concentration after autoclaving, %			
Cotton seeds	1	0,01	0,001	0,01	0,001	0,0001	0,00001
		60	80	65	85	85	80
	2	80	100	75	100	96	100
	3	80	100	80	100	96	100
	4	100	100	80	100	100	100
Cotton seedlings (length, mm)		1,5-24	1,5-24	10-12	20-21	24-26	24-26
Wheat seeds	1	95	85	65	55	80	80
	2	95	100	90	90	100	100
	5	95	100	90	95	100	100
	6	100	100	100	100	100	100
Wheat seedlings (length, mm)		83,6	84,2	71,7	88	94,6	90,6

Influence of Extreme Effects on Properties of BS from *P. Aeruginosa CH5*

Autoclaving

The activity of CB and 1% BS solution of 15-daily culture remained after their autoclaving (121°C, 1 atm.). CB, BS and water-soluble pigments were stable in usual conditions more than 9 months.

Alkaline and Acidic Treatments

Alfalfa seeds were treated with alkaline (pH 8.0-10.0) or subacidic (pH 4.5 - 6.5) solutions of BS. The solutions with pH 10.0 and 4.5 completely suppressed seed germination. The seeds, treated with BS with pH 5.0-8.0, developed normally and, hence BS complex keeps its activity within this pH range.

Treatment with Saline Solutions

BS solutions were prepared with NaCl concentrations 0.85-2.5%. Seeds germinated and actively developed after treatment with BS saline solutions with concentrations 0.85-2.0%, not differening from control.

The data on antagonistic activity of CB and BS towards phytopathogenic fungi are submitted in table 4.

From data of table 4 it is seen, that antagonistic activity of CB and BS is practically equal. The most fungicidal activity the samples display towards to *Verticillium dahliae* phytopathogene.

Table 4. Zones of growth inhibition of phytopathogenic fungi under the influence of CB and BS from *P. aeruginosa CH5*

Fungus culture	Inhibition zone, mm	Inhibition zone, mm
	CB	BS
Verticillium dahliae	75-85	75-87
Rhizoctonia solani	30-40	30-40
Fusarium oxysporum	25-26	26-27

Conclusion

From rhizosphere zone of cotton, alfalfa, wheat and sugar-beet 19 strains of nitrogen-fixing and 9 strains of phosphorous-mobilizing bacteria were isolated, among which two strains originated from wheat rhizosphere were found as active producers of BS. They were identified as *Pseudomonas aeruginosa CH5* and *Bacillus subtilis 26*. On the value of surface tension the optimal nutritive medium was selected, as well as conditions of cultivation in this medium and age of culture (7 days and more), under which the surface tension of CB had the lowest value 36.6 – 40.5 mN/m. From CB of *Ps. aeruginosa CH5* strain, grown up in the medium with addition of sunflower or cotton oil, the complex of lipophilic BS, containing rhamnolipids and phenazine pigments, has been isolated. The content and composition of BS depended on age of culture and nature of vegetative oil, used as source for carbon nutrition. In BS of this strain RL-4 and RL-3 types of rhamnolipids have been identified, in the medium with sunflower oil RL-4 type predominates, in the medium with cotton oil - RL-3 type of rhamnolipid. Autoclaving of CB and BS from *Ps. aeruginosa CH5* can be used as preparations of polyfunctional action for sanitation of sowing material, increase of crop capacity of agricultural cultures, and also as method for bioremediation of sowing lands, contaminated with different pollutions.

Acknowledgements

This research was supported in part in grant No 3200 from the Science and Technology Centre of Ukraine.

References

[1] Vorobiyova L.I. (1989) *Industrial microbiology. M. MSU,* 182-187.

[2] Boronik K.A.M. (2002) Biotechnology of soil remediation on the base of microbial-plant interaction. Proceed. 1st Internation. Congress *"Biology – condition and prospects of development".* M. Russia, 14-18 October, 138.

[3] Gamayurova I.S., Krynitckaya L.Yu., Elizarova Yu.V. (2002) Peculiarity of technology of preparation production for plants protection «Phytospas». Proceed. 1st Internation. Congress *"Biology – condition and prospects of development".* M. Russia, 14-18 October, 137.

[4] Katrukha G.S., Fedorova G.B., Galkina N.N. «Bactofit» – biological fungicide and bactericide of wide spectrum of action. Proceed. 1st Internation. Congress *"Biology – condition and prospects of development".* M. Russia, 14-18 October, 218.

[5] Desai J.D., Danat I.M. (1997) Microbial production of surfactants and their commercial potential. *Microbiol. Molecul. Biol. Reviews.* 61, #1, 47-64.

[6] Gunter N.W. IV, Nunez A., Fett W., Solaiman D.K.Y. (2005) Production of rhamnolipids by *Pseudomonas chlororaphis,* a nonpathogenic bacterium. *Appl. Environ. Microbiol.* 71, #5, 2288 -2293.

[7] Lang S., Wullbrandt D.(1999) Rhamnose lipids – biosynthesis, microbial production and application potential. *Appl. Microbiol. Biotechnol.* 51, 22-32.

[8] Prescott L.M., Harley J.P., Klein D.A. (1993) *Microbiology.* Sec.Ed. Wm.C.Brown Publishers. 912.

[9] Bergey's Manual of Determinative Bacteriology (1997). Ed. D.Hout, N.Kreeg, P. Sneet et al. 9th edition. B.1-2.. M., *«Mir».*

[10] Abramzon A.A. (1981) Surface active substances. L. *«Khimiya»,* 304 p.

[11] Tahzibi A., Fatemeh R., Mahnaz M.A. (2004) Improved production of rhamnolipids by a *Pseudomonas aeruginosa* mutant. *Iran. Biomed. J.* 8(1), 25-31.

[12] Keyts M. (1975) *Technique of lipidology.* Moscow: *«Mir»,* 322.

[13] Shenk T., Schuphan I., Schmidt B. (1995) High-performance liquid chromatographic determination of the rhamnolipids produced by *Pseudomonas aeruginosa. J. Chromatogr.* A 693, 7-13.
[14] Arino S., Marchal R., Vandecasteele J.-P. (1996) Identification and production of a rhamnolipidic biosurfactant by a *Pseudomonas* species. *Appl. Microbiol. Biotechnol.* 45, 162-168.
[15] Hasset D.J., Charniga L., Bean K. et all.(1992) Respons of *Pseudomonas aeruginosa* to pyocyanin: mechanisms of resistance, antioxidant defenses, and demonstration of a manganese-cofactored superoxide dismutase. *Infect.Immun.* 60,#2, 328-336.
[16] Fernandez R.O., Pizzaro R.A. (1997) High-performance liquid chromatographic analysis of *Pseudomonas aeruginosa* phenazines. *J. Chromatogr.* A 771, 99-104.

In: Research Progress in Biotechnology
Editor: G. E. Zaikov, pp. 99-107
ISBN: 978-1-60456-000-8

Chapter XI

Synthesis of Amyloid A-β Fragment with the Use of New Solid-Phase Peptide Synthesis Schedule

M.A. Sanzhakov[1,2], E.Yu. Aleshina* [1], N.V. Pyndyk[1], A.A. Moisa[1] and O.N. Kharybin[1]

[1] V.N. Orekhovich Institute of Biomedical Chemistry, Russian Academy of Medical Sciences, 10, Pogodinskaya ul., 119121 Moscow, Russia

[2] Moscow State Academy of Fine Chemical Technology, 86, Vernadskogo pr., Moscow, Russia

During the solid-phase synthesis of the human β-amyloid fragment, peptide A-β, the influence of several catalysts for 9-fluorenylmethoxycarbonyl protective group removal from α-amino groups of the growing peptide chain, on the yield of target synthesis products and the formation of by-products was studied. The best synthesis results (maximal absolute and relative yields of the target peptide) were achieved when pyrrolidine and 4-methyl-piperidine were used for Fmoc-group removal. The last reagent, being more stable, could both replace piperidine, which is forbidden for a free use, and increase the efficiency of peptide synthesis.

Keywords: solid-phase peptide synthesis, β-amyloid, Fmoc-group removal, 4-methyl-piperidine.

Introduction

Alzheimer's disease is one of the most frequent reasons of dementia in the old and middle age. Etiology of this disease is not clearly ascertained. A principal histological sign of Alzheimer's disease is the accumulation of senile plaques composed mainly of β-amyloid and its fragment, in brain [1, 2]. A hypothesis exists, which is confirmed by a series of

* E.Yu. Aleshina: e-mail: *elena.aleshina@ibmc.msk.ru*; fax +7(495)245-08-57

experiments, that the plaque formation is linked to the isomerization of Asp residues inside the Zn-binding site of β-amyloid, which causes changes in the protein conformation and its aggregation [3, 4]. β-Amyloid fragments may serve as disease biomarkers and be used for the early diagnostics of Alzheimer's disease [5]. Synthesis of β-amyloid fragment three forms – peptide A- β - ^{5}RHDSGY^{10} and its isomers ^{5}RHdSGY^{10} and ^{5}RH*iso*-DSGY^{10} (where d - D-Asp, *iso*-D – *iso*-aspartic acid residue) amidated at their N-termini, was performed to ensure the development of the mass-spectrometric method of the iso-Asp-containing peptide detection in the presence of Asp-containing peptide. Peptides ^{5}RHDSGY^{10} and ^{5}RH*iso*-DSGY^{10} are characterized with a high probability of the side reaction development (L-α-Asp epimerization to *iso*-Asp and piperidide formation) during the stage of 9-fluorenylmethyloxycarbonyl(Fmoc-) protective group removal from α-amino groups of the growing polypeptide chain with the use of a standard procedure applying 20% piperidine solution for Fmoc-group removal [6-8]. The problem of decreasing side reactions is being solved by the use of less basic catalysts of the Fmoc-group elimination than piperidine, such as piperazine and morpholine [7, 8]. However, these reagents show worse efficiency in amino group deblocking [7-9]. The complete removal of Fmoc-groups ensuring the complete attachment of amino acid residues in a peptide is achieved via repeating the reaction several times and/or using more than a fivefold excess of applied reagents [9]. Yet it does not help in increasing the yield of target products in some cases and does not exclude the formation of by-products, according to the experience of certain researchers [7, 9]. Reagent cost and availability should also be considered. In particular, the purchase of piperidine, a standard reagent for Fmoc-group removal, is almost impossible in Russia since 1998, because this reagent is included to the list of controlled reagents (drug precursors) by the Federal Office for the control of the drug turnover. Taking into account piperidine inaccessibility and its inability for ensuring optimal conditions for Fmoc-group removal, this reagent should be replaced. During the synthesis of β-amyloid fragment and its isomers the influence of several reagents for Fmoc-group removal on the yield of target products and by-product formation has been studied.

Materials and Methods

Reagents

N$^{\alpha}$-Fmoc-L-amino acids, N$^{\alpha}$-Fmoc-D-aspartic acid, N$^{\alpha}$-Fmoc-*iso*-aspartic acid, 1-hydroxybenzotriazole (HOBT) were from Novabiochem (Läufelfingen, Switzerland), 1,8-diazabicyclo[5.4.0]undec-7-ene (DBU), piperazine, pyrrolidine (PYR), 4-methyl-piperidine (4-MPIP), morpholine, 3,6-dioxa-1,8-octanedithiol (DODT), trifluoroacetic acid (TFA), N-methyl-pyrrolidinone (NMP), *iso*-propanol (IPA) were from Acros organics (Geel, Belgium), piperidine (PIP), tri(*iso*-propyl)-silane (TIS) were from Fluka (Buchs, Switzerland), anisole, di(*iso*-propyl)carbodiimide (DIPC), dichloromethane (DCM), methyl-*tert*-butyl ether, and acetonitrile for HPLC were from Merck (Darmstadt, Germany).

N,N-Dimethylformamide (DMF) from Reachim (Russia), reagent grade was purified by azeotropic distillation with water and benzene (DMF:water:benzene 125:15:6, v/v) followed by vacuum distillation at 52-53°C / 20 mm Hg [10]. Before use in the peptide synthesis DMF was kept under freshly dried (at 300°C) 0,4-nm molecular sieves (Merck) for at least 3 days, 80 g sieves per 1 l DMF.

Peptide Synthesis

Peptides were synthesized by a manual solid-phase synthesis via the sequential chain elongation starting from the C-terminal residue on D-series Lantern Rink-amide polyethylene pins with 8-mkmol capacity from "Mimotopes" (Clayton, Australia). *tert*-Butyl groups were used for blocking OH side-chains of Ser and Tyr, trityl group was used for imidazolyl group of His, *tert*-butoxy – for Asp carboxyl, and 2,2,4,6,7-pentamethyldihydrobenzofuran-5-sulfonyl (Pbf) for guanidino group of Arg protection. группу, для Asp - OtBu-группу. Carbodiimide method (DIPC in the presence of HOBT in the equimolar ratio to Fmoc-amino acid) was used for Fmoc-amino acid attachments to the growing peptide chain. 5-fold excess of Fmoc-amino acids (in relation to solid phase active group amount) was used throughout the condensation steps. Peptide removal from the polymer together with the side-chain functional group deblocking was carried in 2 ml of the TFA:DODT:TIS:anisole:water mixture (183:5:2:5:5) for 4 hours. The solution was then evaporated under reduced pressure, and the residue was resuspended in 4 ml of methyl-*tert*-butyl ether:petroleum ether (1:2, v/v) mixture containing 0,1% 2-mercaptoethanol. The suspension was chilled to -20°C for 30 min and centrifuged. Supernatant was discarded, the deposit was resuspended in 2 ml of methyl-*tert*-butyl ether:petroleum ether (1:2, v/v) mixture, chilled for 30 min and centrifuged again. The residue was dissolved in 1 ml of acetic acid:acetonitrile:water (1:0,6:10, v/v) mixture. The solution was evaporated under reduced pressure, and the residue was dried in the desiccator under KOH and vacuum for 2 days.

Fmoc-Group Removal

Reagents used for Fmoc-group deblocking:

1. 20% PIP in DMF;
2. 20%PIP+ 2%DBU in DMF;
3. 50% morpholine in DMF;
4. 6% piperazine in DMF;
5. 20%4MPIP in DMF;
6. 20%4MPIP + 2%DBU in DMF;
7. 20%Pyr in DMF;
8. 20%Pyr + 2%DBU in DMF.

Pins were incubated 20 min in one of the above mentioned mixtures. Upon deblocking of α-amino group of Asp (or *iso*-Asp) reagents 2, 6, or 8 were replaced by reagents 1, 5, or 7, correspondingly.

Analysis of Peptide Preparations

Analytic reversed-phase HPLC was performed on the column ZORBAX SB-C18 150x4,6 mm with the help of the Agilent 1100 Series Workstation (Agilent Technologies/Interlab, Moscow, Russia). Peptides were eluted with a linear gradient of buffer A in buffer B from 0% to 100% (for 16 min); buffer A - 0,05% TFA in water, buffer B - 0,05% TFA in acetonitrile, elution speed 1.0 ml/min. Absorption at 214 nm was registered.

Peptide structures were confirmed by matrix-assisted laser desorption-ionization time-of-flight mass spectrometry (MALDI-TOF MS) on the MALDI-TOF mass spectrometer Reflex III (Bruker Daltonik, Bremen, Germany) and Fourier-transformed ion-cyclotron resonance mass spectrometry (FTMS) on the mass spectrometer Apex-Qe FTMS (Bruker Daltonik).

Results and Discussion

At present time the standard procedure for Fmoc-group removal in solid-phase peptide synthesis usually consists in the growing peptide chain treatment by 20% (v/v) solution of PIP in DMF, since Fmoc-group is smoothly removed by primary and secondary amines in polar solvents [6]. However, the more and more increasing variety of peptides synthesized on the solid phase sometimes puts new problems concerning the yield of target products and hence generates the need to develop new synthesis schedules and to search for new reagents. Taking that into account, the influence of different reagents for Fmoc-group removal on the yield of target products and formation of by-products was studied in the solid-phase peptide synthesis of a hexamer fragment of β-amyloid, ^{5}RHDSGY^{10} and its isomers ^{5}RHdSGY^{10} and ^{5}RHiso-DSGY^{10} in the form of peptide amides. This β-amyloid fragment contains the sequence Asp^{7}-Ser^{8}, which is highly prone to the base- or acid-catalyzed L-Asp epimerization to *iso*-Asp. Such epimerization can occasionally occur even *in vivo* in brain tissue and is proposed to be a trigger for the plaque formation in Alzheimer's disease [3]. Peptide ^{5}RHiso-DSGY^{10} can be prone to the reverse epimerization of *iso*-Asp to L-α-Asp, while peptide ^{5}RHdSGY^{10} does not undergo epimerization because of its steric structure unfavorable to this process and thus serves a control. Reagents and their concentrations were chosen among those mentioned in the literature though insufficiently studied in peptide synthesis [7, 9-11]. The only exception was 4MPIP, the use of which in solid-phase peptide synthesis for Fmoc-group removal was not described though it could be deduced from its homology to PIP. Another aim of this study was to find a replacement to PIP, which is a drug precursor and thus a controlled reagent. The whole set of tested Fmoc-removal catalysts included PIP, piperazine, morpholine, Pyr, 4MPIP, and also DBU combined to PIP, Pyr, or 4MPIP (table).

Table. Influence of reagents for Fmoc-group removal on the yield of peptide synthesis and by-product formation

No. #	Reagent for Fmoc-group removal	Peptide synthesized	Crude peptide yield, mg	Asp residue modification*	Epimerization Asp/iso-Asp**	Relative target peptide yield, %
1	20% PIP in DMF	^{5}RHdSGY^{10}	5,3	No	-	31
2	20% PIP in DMF	^{5}RHiso-DSGY^{10}	6,0	No	Yes	20,3
3	20%PIP+ 2%DBU in DMF	^{5}RHDSGY^{10}	6,5	No	No	27
4	20%PIP+ 2%DBU in DMF	^{5}RHdSGY^{10}	7,5	No	-	25
5	20%PIP+2%DBUinDMF	^{5}RHiso-DSGY^{10}	5,8	No	No	20,7
6	50%Morpholine in DMF	^{5}RHdSGY^{10}	5,9	No	-	23,2
7	50%Morpholine in DMF	^{5}RHiso-DSGY^{10}	4,0	No	No	21,3
8	6% Piperazine in DMF	^{5}RHdSGY^{10}	5,2	No	-	25,9
9	6% Piperazine in DMF	^{5}RHiso-DSGY^{10}	4,2	No	Yes	20
10	20%Pyr + 2%DBU in DMF	^{5}RHDSGY^{10}	6,5	Yes	No	48
11	20%Pyr in DMF	^{5}RHDSGY^{10}	7,4	Yes	No	53,9
12	20%4MPIP in DMF	^{5}RHDSGY^{10}	9,3	No	No	50,8
13	20%4MPIP+2%DBU in DMF	^{5}RHDSGY^{10}	9,5	No	No	65

* - Formation of imide between the secondary amine and Asp carboxy-group.

** "- " – means the impossibility of Asp epimerization (in case of D-Asp residue).

Rather low yields of the target products were observed while using standard conditions for Fmoc- removal (20% PIP in DMF, see table and figure, A). According to MALDI-TOF MS analysis shorter peptides (mainly with Tyr deletion) consist a sufficient portion of the peptide samples. It was shown that in the case of the use of 20% PIP in DMF for Fmoc-group removal the epimerization of *iso*-Asp to L-Asp occurred, at least partly. Epimers of target peptides were revealed with the help of a special method of ion-cyclotron resonance MS/MS analysis on FTMS instrument via the formation of the $b+H_2O$ ion by *iso*-Asp-containing peptide [12]. Surprisingly, the use of PIP mixture with DBU did not significantly influenced the yields of target peptides and did not cause Asp epimerization (table and figure, B).

Some authors recommend the use of weaker bases, such as piperazine and morpholine, for Fmoc-group removal as a means of increasing the yield of target peptides containing sequences Asp-Gly, Asp-Ser, Asp-Ala, Asp-Asn, highly prone to Asp epimerization [7-9]. However, in our case piperazine and morpholine application for α-amino-group deblocking led to an even smaller yield of target peptides, and *iso*-Asp underwent epimerization while using piperazine (table and figure, C). The best results for the synthesis of all three peptides (maximal yield of the target peptide) were achieved, when Pyr and 4MPIP were used for Fmoc-group removal (see table and figure, D). Mechanism of Fmoc-group removal is not clearly understood up to now, however it is shown that Fmoc-group elimination goes through the mechanism E1cB [6]. Hydrogen attached to β-carbon atom of the Fmoc-group is labile to bases. Upon the base action on a compound with an amine group blocked with Fmoc an intermediate carbanion is formed that further splits to a free amine and dibenzofulvene, which forms an adduct with bases. The stability of carbanion, formed by means of β-elimination, and the base character define the relative ease of deblocking. $pK_{BH}{}^{+}$ values for PIP, Pyr, and 4-MPIP lie in the narrow interval from 11,27 to 11,12. Such small difference in the basicity does not define any substantial advantage of one base compared to another in the efficiency of Fmoc elimination. One can suppose that 4MPIP, which is a somewhat weaker base than PIP, but a stronger base than piperazine and morpholine, is an optimal reagent ensuring an efficient Fmoc-group removal together with decreasing by-product formation. Pyr ensured easy Fmoc-group removal as well, however, the use of the mixture of Pyr with DBU (in contrast to 4MPIP with DBU) led to the formation of relatively more by-products (including the target peptide with modified Asp residue, Asp-pyrrolidide) (see table). Besides that, Pyr as 20% solution in DMF was quickly oxidized by air oxygen and light that required the everyday preparation of the fresh reagent and showed its inapplicability in automated peptide synthesis.

Hence 4-methylpiperidine represents an optimal catalyst for Fmoc-group removal from α-amino groups in solid-phase peptide synthesis. It can both replace piperidine, a precursor forbidden for free purchase, and increase the efficiency of the solid-phase peptide synthesis. The latter should contribute to the unification of the solid-phase peptide synthesis that is especially important for multiple parallel synthesis of peptide chips.

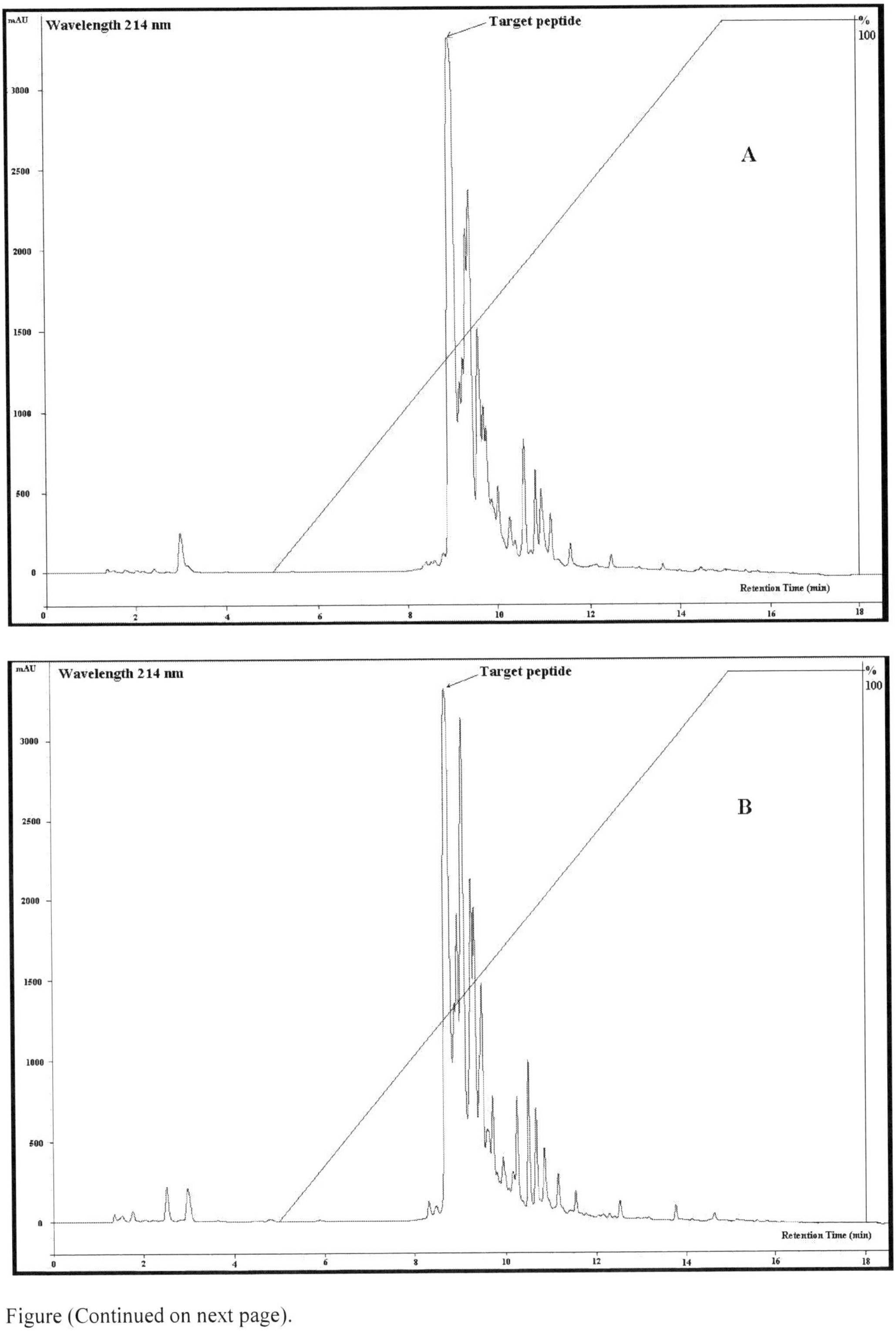

Figure (Continued on next page).

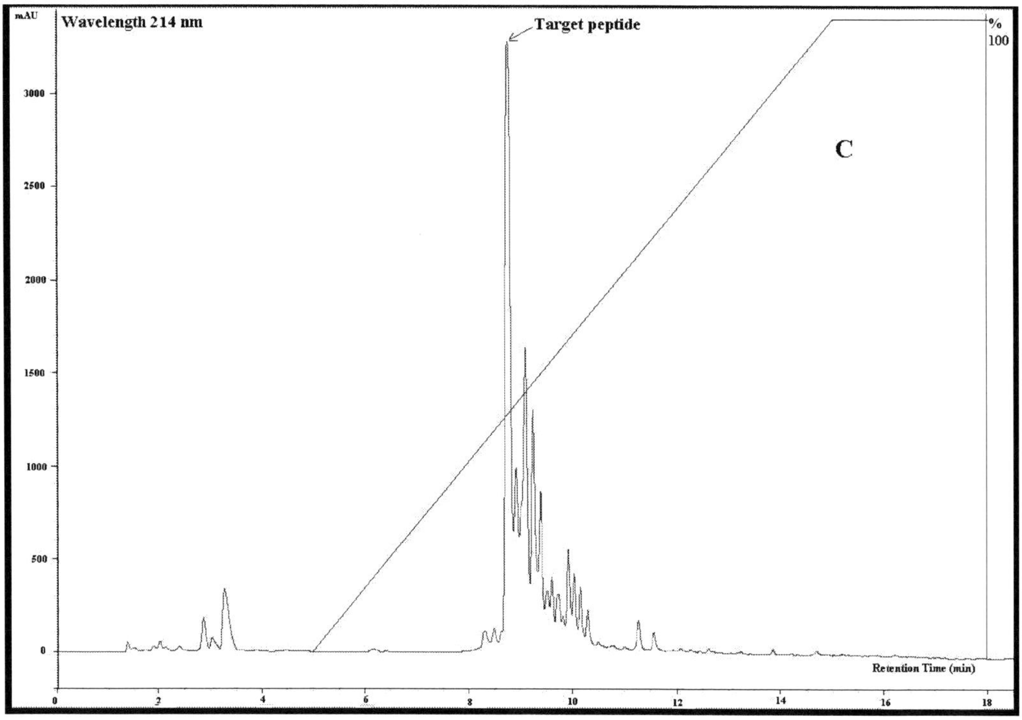

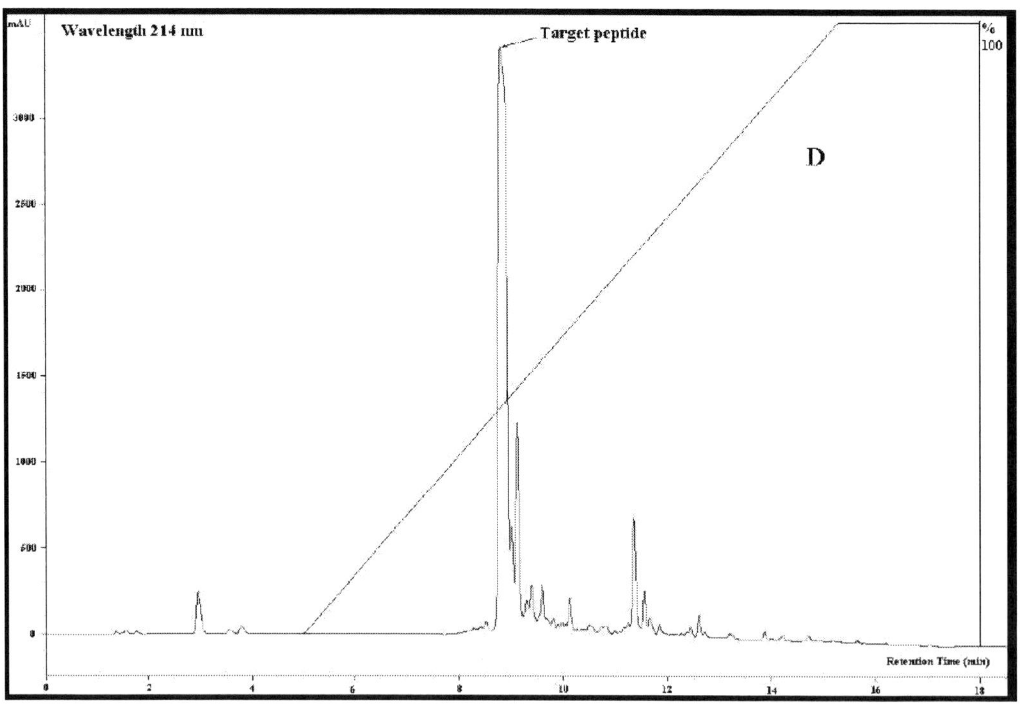

Figure. Results of analytical HPLC of peptide preparations: A – RHDSGY synthesized via using PIP+DBU (3; No. according to table); B – RHiso-DSGY synthesized via using PIP+DBU (5); C – RHdSGY synthesized via using piperazine (8); D - RHDSGY synthesized via using 4MPIP (12). HPLC procedure is described in Materials and Methods.

References

[1] A. B. Clippingdale, J. D. Wade, C .J. Barrow: *J. Pept. Sci.,* 7 (5), 227 (2001).

[2] J. Hardy, D. J. Selkoe: *Science.,* 297 (5580), 353 (2002).

[3] S. Zirah, S.A. Kozin, A.K. Mazur, A. Blond, M. Cheminant, I. Segalas-Milazzo, P. Debey, S. Rebuffat: *J. Biol Chem.,* 281 (4), 2151 (2006).

[4] M. Meyer-Luehmann, J. Coomaraswamy, T. Bolmont, S. Kaeser, C. Schaefer, E. Kilger, A. Neuenschwander, D. Abramowski, P. Frey, A.L. Jaton, J.M. Vigouret, P. Paganetti, D.M. Walsh, P.M. Mathews, J. Ghiso, M. Staufenbiel, L.C. Walker, M. Jucker: *Science,* 313 (5794), 1781 (2006).

[5] D. Galasko: *J. Alzheimers Dis.,* 8 (4), 339 (2005).

[6] P. Lloyd- Williams, F. Albericio, E. Giralt: Ny, *CRC Press Llc,* 1997. 280 P.

[7] J. D. Wade, M.N. Mathieu, M. Macris, W. T. Geoffrey*: Lett. Peptide Sci.,* 2004.

[8] J. Cebrian, V. Domingo, F. *Reig: J. Peptide Res.,* 62, 238 (2003).

[9] B.D. Larsen, A. Holm: *Int. J. Pept. Protein Res.,* 43 (1), 1 (1994).

[10] H. Becker Et Al.: *Organicum,* V.2. *Mir,* Moscow, 1979. 442 P. (In Russian).

[11] G.B. Fields, R.L. Noble: *Int. J. Pept. Protein Res., 35,* 161 (1990).

[12] C. Hyde, T. Johnson, D. Owen, M. Quibell, R.C. Sheppard*: Int. J. Pept. Protein Res.,* 43 (5), 431 (1994).

[13] L.J. Gonzalez, T. Shimizu, Y. Satomi, L. Betancourt, V. Besada, G. Padron, R. Orlando, T. Shirasawa, Y. Shimonishi, T. Takao: *Rapid Commun. Mass Spectrom.,* 14 (22), 2092 (2000).

In: Research Progress in Biotechnology
Editor: G. E. Zaikov, pp. 109-115

ISBN: 978-1-60456-000-8

Chapter XII

New Method of Diagnostics of Possible Genetoxic Effects of Food Products and Medicinal Preparations

V.S. Sibirtsev*[1] and A.V. Garabadzhiu[2]
[1] Mendeleyev Russian Research Institute for Metrology;
Moscowskij pr. 19, St. Petersburg, 190005 Russia
[2] St. Petersburg State Technological Institute,
Moscowskij pr. 26, St. Petersburg, 190013 Russia

Abstracts

New protocol of DNA analysis (*a using system from two fluorescent nucleic-specific dyes with agreed properties*) are considered – which enables to evaluate not only DNA total in a sample, but also change degree of a cell genome structure. Number of new biotesting methods, developed on the basis of above mentioned DNA analysis protocol, are considered also – which enables to evaluate: "general microbial pollution (*seeding*)" of liquid environments; possible genetoxical effects of various food products, medicinal preparations, damaging factors of environment (*including at joint their action on live organismes*); individual radiating sensitivity of organism and etc.

Keywords: biotesting, DNA analysis, fluorescent dyes on DNA.

Recently, as a result of active development of pharmacology and various biotechnology methods, as well as increase of common level of anthropogenic workload on environment, rather impotant becomes the problem is enough fast valuation of opportunity any genetoxic effects from new or modificated food products, medicinal preparations and etc. – and even as

* To whom correspondence should be addressed; e-mail:vs1969r@mail.ru; websyte: http://www.vs1969 r.narod.ru/publen.htm

at isolated action of them on live organism, as in a combination of them to other food products, medicinal preparations, various environment factors, specific for that region, and etc. And besides is frequently important problem is reasonably fast and complex valuation of ecological nonwell-being degree of any region – and even in case, when the standard methods do not there fix excess of the norms on one of possible factors; however in a combination with one another they can make damages on the people and other live organismes.

Above mentioned problems are resolved can be different ways... but the most acceptable use for these purposes standard biosystems (*which under characteristics are close to human organism, are cheap, are accessible in mass quantity and have relatively short life cycle*)... one of major parameters, describing condition of which, are changes in a structure them cell genome... which, in turn, is the most simple and quickly, in our opinion, can be revealed with the help of synthetic fluorophores, capable to interact with DNA including directly in a cell nucleus and sensitive to presence even rather small quantities of nucleic acids... owing to that, DNA-specific fluorophores can be used, besides mentioned, as medicinal preparations, in research, diagnostic purposes and etc.

Pursuant to above said, purpose of the present work has become:

- to develop with the help of DNA-specific fluorophores a method of the analysis not only nucleic acids total, but also changes in a cell genome structure;
- to develop on the basis of above-stated, as well as using standard biosystems, method of valuation of possible genetoxical effects of various food products, medicinal preparations, damaging factors of environment (*including at joint their action on live organismes*);
- as well as to consider opportunities of application of DNA-specific fluorophores in other methods of diagnostics and monitoring, based on the quantitative and qualitative analysis of a genetic material in cells.

DNA Analysis with the Help of System from Two Dyes with Agreed Properties

At all variety of above mentioned potential opportunities for use флуорофоров the spectrum of methods particular, practical application of such substances so far more far from so is great, as it would be necessary. It is result, in particular, that the "strong selective linkage with substrat, as a rule, requires from ligand of one structural features, and availability of properties suitable for registration – other. And one of the best ways of the sanction of this contradiction we consider use for DNA analysis of systems including several nucleotide specific fluorophores, which are indirectly linked together (*as far as the separate fragments of large complex substance can not only mutually supplement, but also to interfere one another*) via the common for them substrate, but share concerted complex forming and spectral properties.

As example of one of such systems by us were chosen two widely used (*on separateness*) commercial DNA–specific fluorophores: Hoechst–33258 *(2-[2-(4-hydroxyphenyl)-5(6)-*

benzimidazoyl]-5(6)-(1-methyl-4-piperazinyl)benzimidazole) (Ht) and ethidium bromide (*2,7-diamino-10-ethyl-9-phenylphenanthridinum bromide*) (EB). First of which (EB) is an intercalator that lies between two base pairs of the polynucleotide double helix; it exhibits mod erate preference to GC-rich sites, but basically specific to secondary and higher orders of DNA molecule organization [1]. Whereas second (Ht) joins a DNA molecule "outside" and is preferablly specific for sites of polynucleotide double helix containing three sequentially positioned AT and only GC base pairs [2] (*see figure 1*). And besides at joint sorption of Ht and EB on a polynucleotide between molecules these dyes a fluorescent resonant non-radiative energy transfer (FRET) can occur.

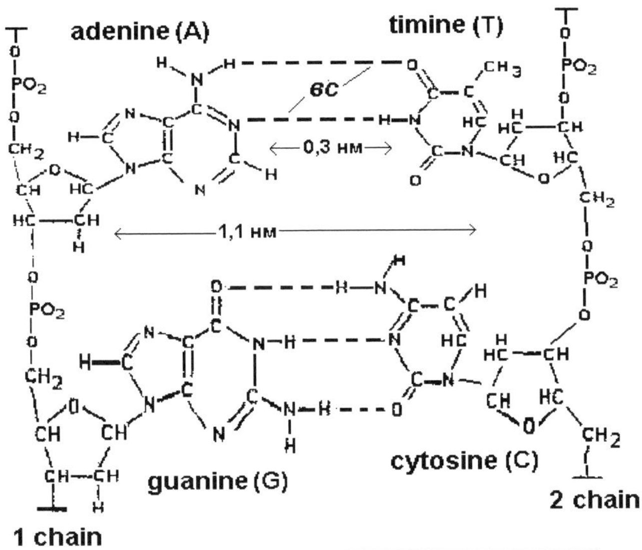

Example of DNA primary structure

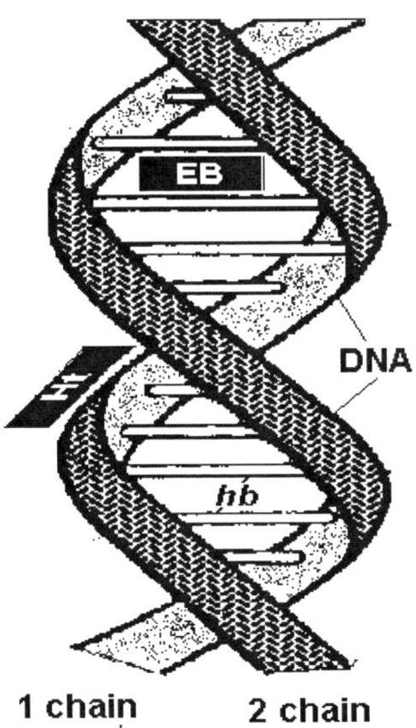

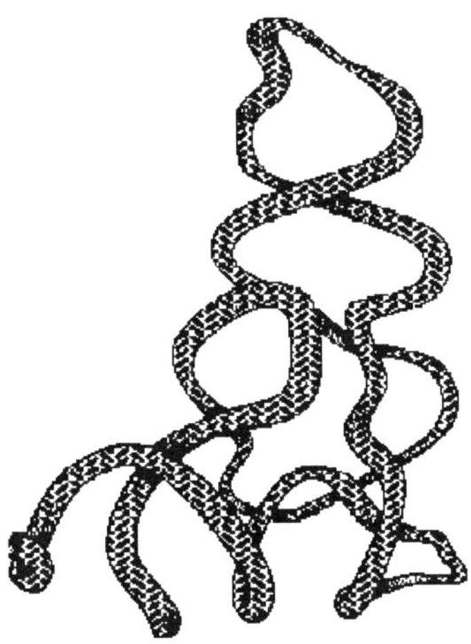

Example of tertiary structure of prokaryote DNA

Figure 1. The schemes of DNA structure ДНК and interaction of intercalators (EB) and nonintercalators (Ht) with DNA. "hb" are designated the hydrogen links between complementary nucleotides of the first and second chains of DNA.

The FRET is the phenomenon, when at fluorescence excitation of energy donor (*in our case this is* Ht) molecules begin to shine not only they, but also energy acceptor (*in our case this is* EB) molecules. FRET can occur only, when the fluorescent emission wavelength region of energy donor is largely crossed with the fluorescence excitation wavelength region of energy acceptor (*as takes place in the case of* Ht *and* EB) (*see figure 2*). And the FRET efficiency rather considerably depends on distance between chromophore groups of the energy donor and acceptor [3,4].

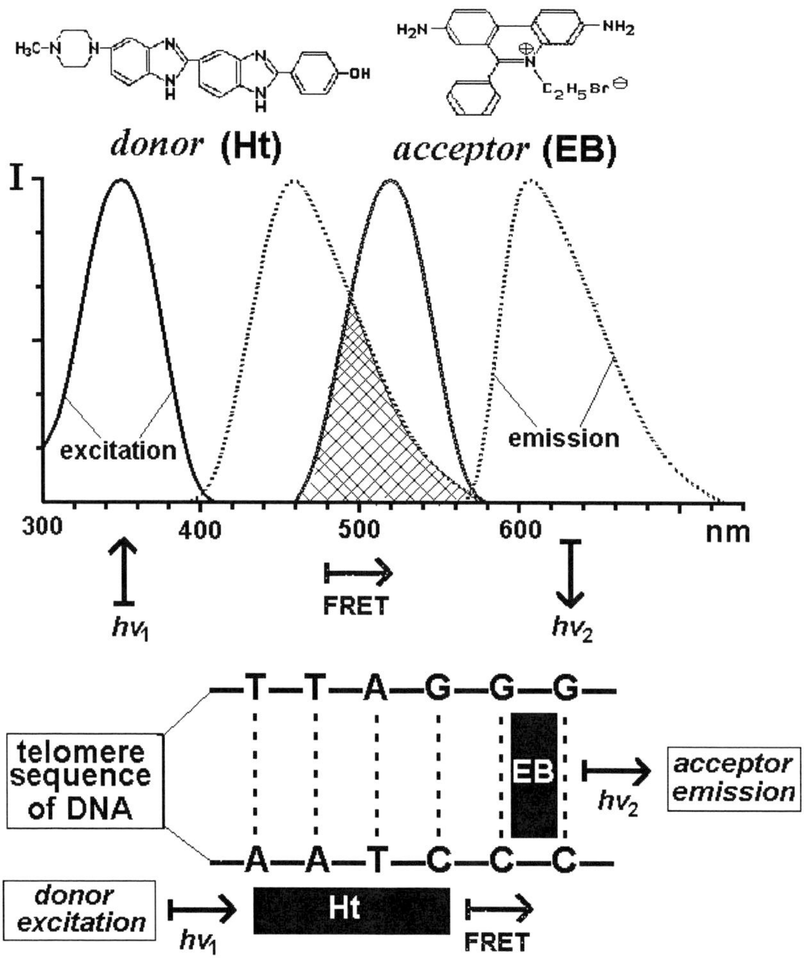

Figure 2. The scheme of fluorescent resonant energy transfer (FRET) between molecules Ht and EB. I are designated the fluorescence intensity. $h\nu_1$ and $h\nu_2$ are designated the light quantums, absorbed Ht and emited EB, accordingly.

Thus obviously, that at the combined use of Ht and EB it is possible to evaluate not only DNA total in a sample, but also change degree of a cell genome structure. Below we consider a practical example of one possible protocol of genetic material analysis using the system "Ht+EB".

1) Mix 0.1ml sample (*e.g., whole blood*) with 1ml of the water lysing mixture – used that to infringe integrity of cell walls, as well as external cell and nuclear membranes, and thus to ensure access of dye from "exter-cellular" solution to "inter-cellular" DNA, and containing 2M *NaCl* + 0.1M *Na_2EDTA* (*ethylenediamine tetraacetate, disodium salt*) + 0.01M *Tris* (*2-amino-2-hydroxymethyl-1,3-propane diol*) + 0.5% (v/v) *Triton X-100* (*4-octyl-[2,4,6,8,10-decapentol]hydroxybenzene*) (pH 8.0). Incubate resultant mixture for 5min. Then, take 0.05ml aliquot and mix it with 1ml of standard water buffer – containing 0.01M *NaCl* + 0.01M *Na_2EDTA* + 0.01M *Tris* (pH 7.4). As a result, we receive a solution 1, for which we measure background fluorescence intensities: I_{FH} (a*t excitation and emission wavelengths: 350 and 455 nm – appropriate to complex of* Ht *with DNA*) and I_{FE} (a*t excitation and emission wavelengths: 520 and 605 nm – appropriate to complex of* EB *with DNA*).

2) Add 0.05ml of Ht dissolved in a standard water buffer in concentration 10μg/ml to 1.05ml of solution 1. As a result, we receive a solution 2, for which at excitation and emission wavelengths 350 and 455 nm we measure fluorescence intensity I_H. Then, add 0.02ml of standard DNA dissolved in a standard water buffer in concentration 60μg/ml to 1.1ml of solution 2. As a result, we receive a solution 3, for which at the same wavelengths, as earlier we measure fluorescence intensity I_{DH}. Then, we calculate a DNA total in a sample as: $C_{DH}=g \cdot C_{DS} \cdot (I_H - I_{FH})/(I_{DH} - I_H)$ – where g=246 is analysed DNA dilution in solution 3 compared with the whole blood, and CDS=$3.3 \cdot 10^{-6}$M is concentration of standard DNA in solution 3.

3) Add 0.05ml of EB dissolved in a standard water buffer in concentration 200μg/ml to 1.05ml of solution 1. As a result, we receive a solution 4, for which at excitation and emission wavelengths 520 and 605 nm we measure fluorescence intensity I_E. Then, add 0.02ml of standard DNA dissolved in a standard water buffer in concentration 60μg/ml to 1.1ml of solution 4. As a result, we receive a solution 5, for which at the same wavelengths, as earlier we measure fluorescence intensity I_{DE}. Then, we calculate a structural factor: $Ks=C_{DE}/C_{DH}$ (*where* $C_{DE}=g \cdot C_{DS} \cdot [I_E - I_{FE}]/[I_{DE} - I_E]$) – change of significance of which should, in particular, to reflect changes in character of the tertiary and higher orders of organization of genome structure in a sample (as far as EB, as was already marked, to it is essentially more specific, than Ht).

4) At first, we measure I_{FHE} – fluorescence intensity of solution 2 at excitation and emission wavelengths 520 and 605 nm. Then, add 0.02ml of EB dissolved in a standard water buffer in concentration 200μg/ml to 1.12ml of solution 2. As a result, we receive a solution 6, for which at excitation and emission wavelengths 520 and 605 nm we measure fluorescence intensity I_{HE}. Then, we calculate a factor of effectiveness of energy transfer: $Kt=(I_{HE} - I_{FHE})/C_{DH}$ – the reduction of significance of which should, in particular, as shown in fig.2, in a certain measure to reflect reduction of a length of genome telomere sites... which for all vertebrates, for example, represent fragments "TTAGGG" repeated on the ends of DNA molecules... the quantity of which decreases at each cell division (*due to DNA replication problem*), and accordingly determines the maximal limit of live organism age... executing also some other, in particular regulating, function [5,6].

Valuation of Genetoxic Effects of Various Products, Preparations and Environment Factors

Being based on above-stated us were developed: method for valuation of possible genetoxical effects of various food products and medicinal preparations on live organismes (*including at joint action on these organismes of other products, preparations or environment factors, specific for any region*)... and method for valuation of possible genetoxical effects of low dozes radiating, chemical, microbiological and other damaging factors of environment att joint their action on live organismes. These methods are based that group of animals (*for example, rats, the life cycle of which proceeds about in 20 times faster, than at the people*) during a certain time feed with a analyzed product (*or enter to him a appropriate preparation, or subject to effect of damaging factor*)... in a combination, if necessary, with other products, preparations or environment factors. At it samples of blood from animals regularly select. Then process by it special lysing mixture (*its contents see above*). Then evaluate a DNA total and genome structure in sample by special dye (*or system from several dyes – such as* Ht *and* EB, *for example*)... ration them on quantity of leucocytes (*as far as erythrocytes in a normal condition in mammal blood do not contain of DNA*)... compare received parameters to similar values, determined for control animal group, as well as for all animals, participating in analysis, prior to the beginning its realization. And on the basis of this comparison come to the conclusion about availability of any genetoxic effect of analyzed products or factors on live organisms.

Valuation of "Microbial Pollution (*Seeding*)" of Liquid Environments

The new method for express valuation of "general microbial pollution (seeding)" of liquid environments (which is one of important parameters at valuation of quality for a water and food products, at the monitoring of various biotechnological manufactures and etc.) was developed us also. Standard methods, used for this purpose at present (and including: seeding of analyzed material on medium; incubation it then at constant temperature, in steril conditions during from one till several days and subsequent count of number of formed microbial colones), require for realization of a significant time and specialized stationary conditions. And our method permits to execute the similar analysis practically in a "real time mode" and directly on those places, where samples were taken, and consists in following.

Mix 1ml analysed sample with 0.1ml of the water lysing mixture (*its contents see above*). Then received solution we agitate, incubate during several minutes (*at* $t^{O}=25\pm3^{O}C$ – *up to establishment of stable indications on fluorimeter*). Then we measure background fluorescence intensity of sample I_{F1}. Then to analyzed solution we add also 0,1ml solution (*the background fluorescence intensity which* I_{F2} *is defined beforehand*), containing by 10□g/ml Ht (*or other DNA-specific fluorophor in other concentration*) in a standard water buffer (*its contents see above*). And as well as earlier, we determine new fluorescence intensity I_D. Then, we add also 0.02ml of standard DNA (*for example, thymus DNA dissolved in a standard water buffer in concentration 60μg/ml*) to analyzed solution. And as well as earlier, we determine new fluorescence intensity I_{DS}. Then we calculate microbe total in a sample as: $Cm=a_0+a_1(I_D-I_{F1}-I_{F2})/(I_{DS}-I_D)$ – where empirical factors a_0 and a_1 are defined on a data of calibration with use standards with the known microbe contents.

This method (*on which we received the patent [13]*), using a TKO-100 portable microfluorimeter, has allowed to define microbe total in sample since 20cl/ml, with error not more than 30%, expending not more than 10min on one measurement. Besides in particular, if instead of lysing mixture in above-described method to use any antibiotic, forming channel only in external wall and membrane of cell – DNA-specific fluorophor will paint probablly only prokaryotes. And if to deduct these data from microbe total in a sample (*determined as is described above*), it will be possible to evaluate also quantity of eukaryotes, contained in analyzed sample. And on manufactures, realized in stationary conditions, for registration of staining of DNA it is possible to use besides already not portable microfluorimeter and stationary flow cytofluorimeter – that will allow not only to execute valuation of "microbial pollution" really in a "real time mode", but also essentially to increase sensitivity of such valuation.

Valuation of Individual Radiating Sensitivity of Organism

At last, we developed a method of valuation of individual radiating sensitivity of animals and people, based that at cancer diseased, for example, prior to the beginning radiation therapy take sample of blood (*it is enough 1ml*), adulterate by a standard buffer (*its contents see above*) and divide on two parts, one of which then subject to action of X-ray and other (*control*) – leave as is. Then each of these samples incubates during 3 hours at 37^0C (*for reparation of DNA primary damages*) and process then by a lysing mixture. Then valuation of DNA total and genome structure in a sample by special dye (*or system from several dyes – as for example* Ht *and* EB) and fluorimeter... and ration received values on quantity of leucocytes. At last, comparing received values among themselves, judge a degree of damaging effect, which will render probablly radiation therapy in a chosen doze on this particular diseased.

More information on our works in it and other adjacent directions can be on websyte: *http://www.vs1969r.narod.ru/publen.htm*.

References

[1] A.R.Morgan, J.S.Lee, D.E.Pulleyblank, N.L.Murray, D.H.Evans: *Nucl. Acids Res.* **7**, 547–571 (1979)

[2] P.E.Pjura, K.Grzeskowiuk, R.E.Dickerson: *J. Mol. Biol.* 197, 257–271 (1987)

[3] V.S.Sibirtsev: *Biochemistry.* (Moscow). 70 (4), 449–457 (2005)

[4] S.Speiser: *Chem. Rev.* 96, 1953–1976 (1996)

[5] A.K.Miker, D.S.Koffi: *Biochemistry.* (Moscow). 62 (11), 1547–1557 (1997)

[6] E.X.Blekbern: *Biochemistry*. (Moscow). 62 (11), 1400–1406 (1997)

[7] S.D.Ivanov, V.S.Sibirtsev: Method of valuation general microbiological pollution. Russia patent №2079138, bulletin №13 (1997).

In: Research Progress in Biotechnology
Editor: G. E. Zaikov, pp. 117-124
ISBN: 978-1-60456-000-8

Chapter XIII

Adsorbent for Purification of Gas-Air Mixture. Humoral Immunity and Energy Status of an Organism

I.V. Zhigacheva[1], E.B. Burlakova[1], L.S. Evseenko[1] and M.G. Voronkov[2]

[1] Institute of Biochemical Physics, Russian Academy of Sciences ul. Kosygina 4,Moscow, Russia

[2] Irkutsk Institute of Organic Chemistry, Russian Academy of Sciences, Irkutsk, Russia

Abstract

The adsorption–catalysis method has been extensively used in industry for adsorption of organic and inorganic compounds. However, after purification of exhaust gases by filtering they still contain volatile nitrosoamines (NA) and polycyclic aromatic hydrocarbons (PAH) in quantities, which, upon chronic exposure, may initiate various pathological processes. At present, a zeolite filter is one of the cheapest and efficient filters used. The aim of this work is to increase the efficiency of adsorption of PAH and NA by zeolite filters. The adsorptive properties of the filters may be improved by impregnating zeolite with an Oncosorb adsorbent – (N, N'-bis (3-triethoxysilylpropyl)-thiocarbamide). The efficiency of adsorption of volatile NA increases from 61.8% to 80% for N-nitrosodimethylamine and from 36.8% to 78.9% for N-nitrorosopiperidine. The efficiency of adsorption of PAH increases from 72% to 85% for benzopyrene, from 36% to 78% for perylene, etc. Enhancement the efficiency of adsorption of toxicants from gas–air mixtures affects biological systems.

Keywords: polycyclic aromatic hydrocarbons (PAHs), nitrosamines (NAs), lipid peroxide oxidation (LPO), leukocyte, liver mitochondria, ά-glycerophosphate dehydrogenase (ά-GPDG), succinate dehydrogenase (SDG).

Numerous polycyclic aromatic hydrocarbons (PAHs), nitrosamines (NAs), and heavy metals are contained in the air of industrial zones, near highways with intense traffic, and indoors where tobacco is smoked. These substances are accumulated in soil, water, and food [1]. After a long exposure, even microscopic doses of them lead to a number of pathologies [2]. At present, a zeolite filter is one of the cheapest and efficient filters used. The aim of this work is to increase the efficiency of adsorption of PAH and NA by zeolite filters. We treated a zeolite by N,N'-bis(3-triethoxysilylpropylthiocarbamide) ("Oncosorb"), possessing effective sorbtion properties [3]. Under the influence of toxicants on the organism, primarily the energy metabolism of the organs and tissues is changed [4]. This permits an estimation of the toxicity of components of gas-air mixtures according to the change in the energy status of the organism. Since toxicants diffuse from the alveola of the lungs into the blood and are detoxified in the liver, a biological evaluation of filter efficiency was conducted by studying the functional state of liver mitochondria and the energy enzymes of lymphocytes ά-Glycerophosphate dehydrogenasae (a-GPDH) and Succinate dehydrogenase (SDH)]. As been shown, a long-term exposure to PAHs and NAs resulted in a decrease in humoral immunity [5]. It was conceivable that these changes were caused by activated lipid peroxiation (LPO) resulting in modification of leukocyte membranes. Therefore, we also studied the correlation between activated lipid peroxide oxidation and changes in humoral immunity.

Materials and Methods

Male Wistar rats with an initial body weight of 90-100 g were used in the experiments. The animals were divided into three groups: (1) control animals; (2) rats exposed to artificial gas-air mixture; (3) rats exposed to artificial gas-air mixture pumped through a filter consisting of a zeolite impregnated with an organosilicon compound Oncosorb. Based on the composition of the air near highways and the indoors air containing tobacco smoke, a special gas-air mixture was prepared (table 1).

Table 1. Composition of the artificial gas-air mixture (ng/m^3)

PAH	Amount	NA	Amount	Metals	Amount
Pyrene	54.0 ± 9.0	NDMA	89.0 ±3.1	Ni	3.0 + 0.5
Benzo(a)pyrene	55.0 ±10.0	NDEA	180.2 ±28.8	Cd	0.5 + 0.1
.Dibenz(a,c)anthracene	12.0 + 3.0	NPYR	175.0+10.8	Mn	0.4 ±0.1
Benzo(a)pyrene	63.0 ±5.0	NPIP	76.0 ±12.0	Pb	0.4 ±0.1
Dibenz(g,h,i) perrilene	50.0+10.0	-	-	Co	1.0 ±0.5
.Dibenz(a,h)antracene anthracene	10.0 ±2.0	-	-	Mg	5.0 ±0.8

Note: NDMA, N-nitrosodimethylamine; NDEA, N-nitrosodiethylamine; NPYR, N-nitrosopyrrolidine; NPIP, N-nitrosopiperidine.

The animals, two or three at a time, were exposed to this gas-air mixture in a 5-liters desiccator; the mixture entered the desiccator by means of an underpressure created by a vacuum pump. The animals were exposed to the gas-air mixture for 1.5 min at 30 to 40 min intervals, four times a day.

The content of PAHs and volatile NA determinated according methods described earlier [6]

Liver mitochondria were isolated according to the method described in [7], The rate of oxygen consumption was recorded with a covered Clark-type electrode on an LP-7E _polarograph at 28°C. The incubation medium contained: 0.2 M sucrose, 10 mM Tris-HCl, 2 mM -KH2PO4, 5 mM MgSO4, and 10 mM KC1, pH 7.5.

The level of LPO was estimated from the LPO product fluorescence [8]. Leukocyte enzymes were detected according to Narcissov [9], and differential blood count was determined upon morphological analysis of the blood samples [10]. In rat spleens, antibody formation was estimated by the Cunningham's method [11] three to six months after exposure to the gas-air mixture. For this purpose, rats were immunized with sheep red cells; after five days, the total number of cells an of antibody-forming cells (AFCs) per spleen determined. In homogenized samples of thymus the content of RNA and DNA was estimate as described in [12]. The prolipheration of lymphocytes according methods described earlier [5], using ^{3}H-thymidine .

Results and Discussion

Passing the artificial gas-air mixtures through a 25mm layer of adsorbent (Oncosorb + zeolite) with rate 19 liters/ hour results in a decrease in a total content of PAHs from 520 ng/m3 to 288,4 ng/m3 . Moreover, the total content of volatile NAs decreased from 244 ng/m3 to 137 ng/m3. In case the using zeolite as an absorbent, the total content of PAHs decreased to 311.5 ng/m3 and the total content of volatile NA decreased to 152.5 ng/m3.

Exposure of animals to the artificial gas-air mixture affects the dynamic of POL processes in various organs and tissues. In the lung tissue and blood plasma, LPO dynamics were somewhat different. In the lung tissue the level of LPO-product fluorescence 1.5 times was increased after two-week exposure. After exposure month, low LPO activation was observed in lung tissue. Three to six months after exposure to the artificial gas-air mixture, a pool of LPO-product was increased (figure 1). In blood plasma, there were not noticeable changes in the level of LPO products early stages of the experiment. However, beginning from the second month, the LPO rate increased, to six months from the beginning of the experiment level of LPO was 1.5- to 2-fold higher than in the control (figure 1). For the liver tissue, the situation was different. Long-term exposure results in two-fold decrease in the level of fluorescence of POL products. It is obvious that differences in the POL dynamic in tissues of liver, lung, and blood are associated with changes in metabolism of ascorbate in the liver tissue; its content increases by 20%. Soon after the exposure to a nitrosamines (NDMA), LPO activation and changes in physical-chemical properties of biological membranes observed in kidney tissue [13]. Three months after beginning of the toxicant exposure, the level of restored to the normal one. The LPO kinetics observed in this study differed from that described in literature probably, because of the combined effect of PAHs, NAs and heavy metals.

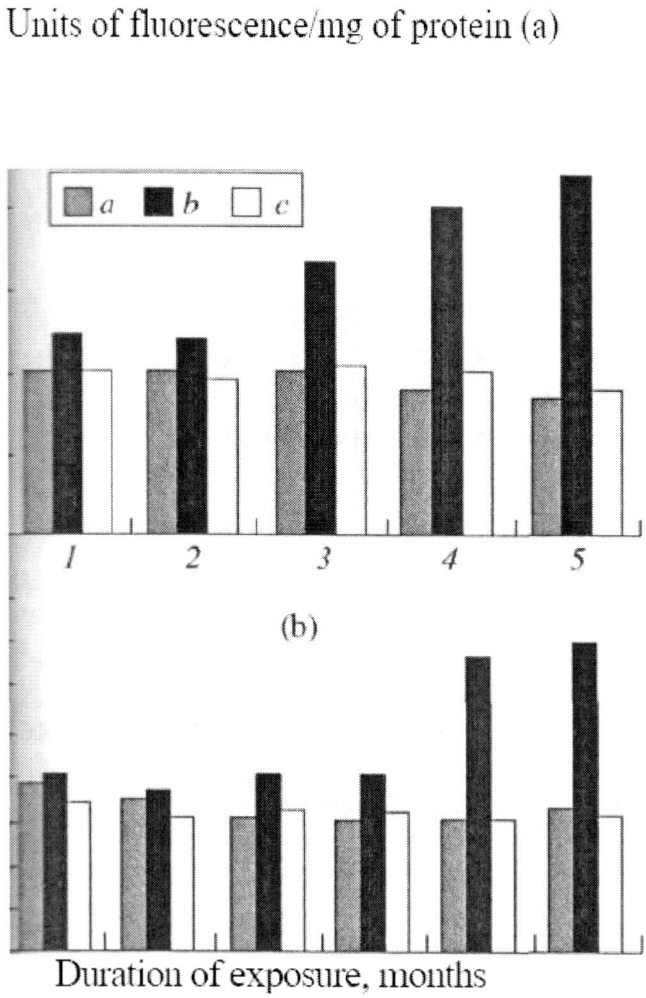

Figure 1. Changes in a pool of POL products in (a) the lung tissue -in (b) blood plasma. Duration of exposure in (a): 1-0,5; *2*.- 1; 3 -*1*, 4-*2*; 5-*6* months. *a*- control; *b*- experiment; *c*-exposure to a filtered mixture.

Along with the above changes observed in plasma, lymphocytic enzymes, such as ά-Glycerophosphate dehydrogenase (ά-GPDG) and Succinate dehydrogenase (SDG), were inhibited. The activity ά-GPDG decreased by 38% as soon as two week from the beginning of the experiment. After two months, this enzymatic activity was 2.2 times below than in the control and after six month, the activity of this enzyme remains lower by 25% as compared with control. (Figure 2a).

The SDG activity decreased dramatically as soon as two weeks from the beginning of the experiments (in 3.3 times). Then it gradual increased, though SDS activity remained 2.1 times lower than control even three to six months afterwards.

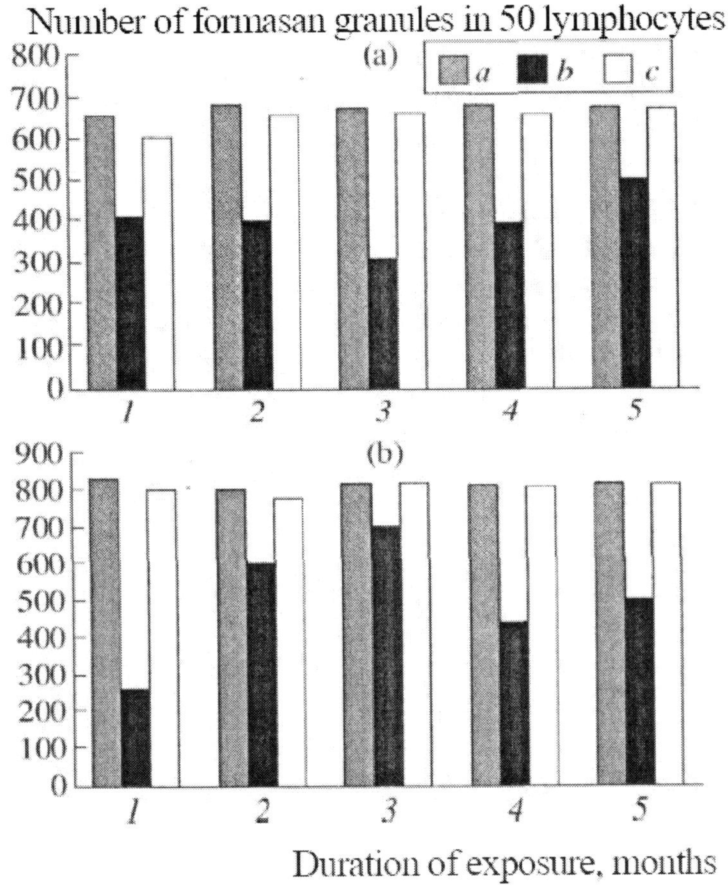

Figure 2. Changes in the activities of (a) a-GPDG and (b) SDG of rat lymphocytes; denotations *a, b,* and *c* are the same as in figure 1.

Changes in differential blood count was another consequence of the exposure to the gas-air mixture, proportion of eosinophils and neutrophils decreased progressively. This was accompanied by a 2-fold increase in the activity of neutrophil acid phosphatase, which is presumably located in lysosomes. Similar changes in lysosome activity of liver endothelioliocytes were induced by benzo(a)pyrene [14].The above changes might have been a consequence of in a tension reaction and increased secretion of corticosteroids [16]. The changes described in energetics of peripheral blood lymphosytes was comparable with those observed in rat liver mitochondria.

In one month of the exposure of artificial gas-air, the maximum rates of oxidation of NAD (nicotineamideadenine dinucleotide)-dependent substrates by respiratory chains of liver mitochondria decreased by 50%. . Thus, in the case of oxidation 4 mM glutamate and 1 mM malate in the presence of 20 mkM 2,4-dinitrophenol (DNP), the rates decrease from 25.8 ± 1.6 nmoles 02/mg mitochondrial protein per min to 11,8 ± 0.4 nmoles 02/mg mitochondrial protein per min. The efficiency of oxidation phosphorylation is also reduced. Respiratory control according to Chance drops from 2.8±0.3 to 2.0±0.2. However if succinate is used as the oxidation substrate, then the differences between the control and experimental animals disappear. After two months maximum rates of oxidation of NAD-dependent substrates return to normal. The maximum rates of oxidation of succinate are decreased by 30%. The rates of succinate oxidation in the presence of DNP drop from 20.0 ± 1.4 to 14.6 ± 1.2 nmoles 02/mg mitochondrial protein per min The degree of coupling of oxidation and phosphorylation also drops under these conditions decreases. The respiratory control according to Chance decreases from 1.8 ± 0. 2 to 1.5 ± 0.2 .Described characteristics remained on this level until the end of experiment.

It is evident, that long-term exposure to gas-air mixture results in a decrease in the activity of enzymes responsible for the energy balance of cells and those effects of toxicants may leads to the disappearance of the adaptive potentialities of the organism. Loss of the activity of enzymes involved in energy metabolism may decrease the functional activity of leukocytes and immune status of the organism. This suggestion was supported by the low number of cells in the spleen, which is a hemopoietic organ. Both the total cell count and that of AFCs were 1.6-2.5 times decreased (table 2). It is believed that the reduced activity of lymphocyte enzymes (ά-GPDG and SDG), disrupted in energy metabolism of liver tissue and decreased number of spleen cells (primarily, AFCs) are related to activation of LPO. The LPO activation may be the main factor leading to a decrease in humoral immunity.

Note that, in response to the toxic effect, the contents of DNA and RNA were increased in thymus tissue. As soon as two months after the beginning of the exposure, the content of RNA was 1.5-fold increased in this organ. The content of DNA was increased a month after the exposure to the gas-air mixture. At that stage, the level of DNA was higher than in the control by 20%; six month from the beginning of the experiment, the increase in the content of DNA reached 40%. The high levels of DNA and RNA syntheses might result from an increase in either nucleotide exchange or protein synthesis. Note that similar changes were observed in the liver after exposure to PAHs, both protein synthesis and mitotic processes were activated in this tissue [15, 16]. An increase in the content of nucleic acids in the thymus may reflect stimulation of proliferation. Indeed, after exposure to the artificial gas-air mixture, 3H-thymidine incorporation into mononuclear cells of peripheral blood was stimulated. After the exposure for three to six months, the rate of lymphocyte proliferation was three times higher than in the control (figure 3). Along with activation of acid phosphatase in leukocytes, stimulation of proliferation in thymus was probably a compensatory process supporting the immune status. Note that the exposure to the gas-air mixture filtered through a zeolite impregnated with Oncosorb caused no changes in the parameters studied. None of these parameters differed from the same parameter in the control group of animals, which suggests that the experimental filter absorbed almost completely the toxicants contained in the gas-air mixture to protect animal organism from their adverse effect.

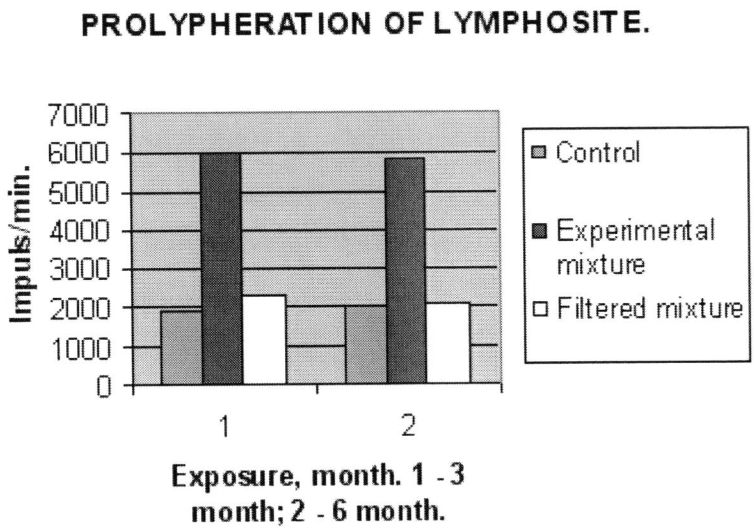

Figure 3. Enfluence of an artificial gas-air mixture on prolypheration of lymphosits.

Table 2. Effects of three- and six-month exposures of rats to the artificial gas-air mixture on the total number of cell and the number of antibody-forming cells in spleen

Group of animals	Number of cells	Number of AFCs per spleen (x108)
Three month exposure		
Control	233100+6720	5.50+0.6
Artificial gas-mixture	84000 +1600	3.36+1.1
Exposure to a filtered mixture	231500+5700	6.00+1.5
Six month exposure		
Control	200600+5670	6.50+1.2
Artificial gas-mixture	55670+1870	2.20+0.9
Exposure to a filtered mixture	200500+6000	7.20+0.5

References

[1] V.N. Gorbunova , in: *Mutageni I kancerogeni v okrugaushey srede Novye podkhody v otsenke riska zdorov'ya* (Environmental Mutagens and Carcii New Approaches to Estimating the Risk for HEALTH, St. Petersburg, 1998, p. 34-58.

[2] B.J. Halliwell, M.C. Gutterige , The Role of transition Metals in Superoxide-Mediated Toxicity: Superoxide Dismutase. Acad. Press, 1985, vol. 3, p. 46-69.

[3] M.G. Voronkov, N.N. Vlasova, Yu. N. Pozhidaev, *Et.al. Dokl. Akad. Nauk SSSR,* 320, No. 3, p. 658-662 .

[4] Yu.S. ROTTENBERG. Author's Abstract of Doctoral Dissertation [in Russian], *Nil po BIKhS,* Moscow ,1986.

[5] I.V. Zhigacheva, E.B. Burlakova, L.S. Evseenko, M.F. Nikonova, E.G. Bulanova, M. Litvina, M.G. Voronkov, *Dokl. Akad. Nauk,* 1999, 367 N 5, P. 705-707. (In Russian)

[6] I.V. Zhigacheva, E.B. Burlakova, L.S. Evseenko, L.E. Krivosheeva, M.I. Savina, M.G. Voronkov, *Dokl. Akad. Nauk,* 1999, 367, N 1, P.117-119. (In Russian)

[7] I.V. Zhigacheva, E.N. Mokhova, V.P. Skulachev, *Dokl. Akad. Nauk Sssr,* 1976, 227, N 2, P. 493-496. (In Russian)

[8] B.I. Flatcher, C.J. Dillard, A.L. Tappel, *Annual. Biochem.,*1973, 1, P. 1-9.

[9] R.P. Nartsissov Arkh. Anat. Gistol. Embriologia, 1969l. 5, P. 85-89.

[10] R.P. Zolotniskaya, *Lab. Delo,* 1983, N 5, P. 36-38.

[11] Dzh.B. Natvich, P. Perlmann, H. Vigzel . *Lymphocytes: Isolation, Fractionation, and Characterization,* Moscow: Meditsyna, 1980.

[12] M.N. Baranov, *Biochimiya,* 1962, 27, N 4, P. 702-705. (In Russian)

[13] V.N. Kirilov, Physical and Chemical Property and Peroxidation of Lipids Of Kidney Subcellulars Structures During Nitrosoamine Carcinogenesis Administration of Antioxidants, Doctoral (Med.) , Chelyabinsk, 1983.

[14] S.V. Michurina, Liver and Its Regional Lymphatic Packs in the Cases of 3, 4-Benzpyrene Exposure during ontogenethis, Cand. Sci. (Biol.) Dissertation, Novosibirsk, 1995.

[15] I.V. Simak, Effects of Carcinogenic Nitrosamines Their Noncarcinogenic Precursors on the Biotransformation System in Rat Liver, Cand. Dissertation, Minsk, 1995.

[16] V.A. Koblyakov in "Environmental Mutagens and Cncerogens "New Approaches to Estimating the Risk, St. Petersburg, 1998, pp. 74-82.

In: Research Progress in Biotechnology
Editor: G. E. Zaikov, pp. 125-130
ISBN: 978-1-60456-000-8

Chapter XIV

Effect of Low Moisture Content on Structural and Functional Characteristics of Plant and Animal Membranes

I.V. Zhigacheva[*1], *L.D. Fatkullina*[1], *A.N. Goloshchapov*[1], *I.F. Rusina*[2] *and E.B. Burlakova*[1]
[1] Emanuel Institute of Biochemical Physics, Russian Academy of Sciences, Moscow, 119334 Russia
[2] Semenov Institute of Chemical Physics, Russian Academy of Sciences, Moscow, 119334 Russia

Abstract

We studied the effects of low moisture and a plant growth regulator – melaphen (melamine salt of bis(oxymethyl)phosphine acid) on the electron transfer rate in the respiratory chain of mitochondria of Pisum sativum pea sprouts. Under these conditions, the maximum rates of oxidation of NAD-dependent substrates by pea sprouts mitochondria decrease by 38% and the electron transfer rates at the end of the respiratory chain decrease by 40%. A preliminary treatment of pea seeds with a 10^{-7} % solution of melaphen results in leveling the difference in the energies of mitochondria of sprouts growing under standard conditions and under conditions of low moisture. The incubation of mitochondria with $4x10^{-12}$ M of melaphen results in a 10% decrease in the microviscosity of the membrane lipid bilayer; the effect is associated with a change in the activity of enzymes of the mitochondrial respiratory chain and, hence, with the antistress efficiency of the compound.

Keywords: plant growth regulators, antiradical activity, mitochondria, low moisture content.

[*] I.V. Zhigacheva: Moscow, 119334 Russia; ul. Kosygina 4, fax: 137-41-01; e-mail: zhigacheva@mail.ru

Apart from specific reactions that depend on particular impacts, nonspecific cellular reactions induced by stressing factors affect the resistance of plant and animal cells to the action of unfavorable environmental factors. Such nonspecific reactions are associated with changes in physicochemical properties of biological membranes. Therefore, it is safe to suppose that agents that can affect the structural characteristics of biological membranes may exhibit adaptogenous properties. As an example of such an agent, a plant growth regulator – melaphen (melamine salt of bis-(oxymethyl)phosphine acid) synthesized at the Arbuzov Institute of Organic and Physical Chemistry, Kazan Research Center, Russian Academy of Sciences was chosen. As a stressing factor applied, Pisum sativum pea seeds were let germinate under low moisture conditions (drought). Melaphen was chosen as an object for study because plant growth regulators are extensively used in agriculture. Investigation of mechanisms of their action is of great importance for economics. Treatment of seeds with plant growth regulators makes it possible to achieve shifts in the plant metabolism induced by certain external factors (temperature, light-time day duration, etc.). These agents increase the resistance of plants to abiotic stresses [1-5]. In particular, melaphen is highly efficient: seeds of a wide gamut of agricultural plants (cereals, legumes, solana) preatreated with melaphen at a concentration of 10^{-7} to 10^{-8} % before sowing results in a 5 to 25 % increase in the energy of germination and productivity, and a change in the amino acid composition of wheat grain proteins [6,7]. In addition, melaphen used at a concentration of 3×10^{-10} M increases the biosynthesis and plant cells respiration rates by 15% with energy losses remaining on the control level [8]. Supposedly, these effects result from changes in the structural characteristics of membranes and, hence, in the activity of membrane-associated enzymes of the mitochondrial respiratory chain and electron-transport chain.

Materials and Methods

The experiments were performed in mitochondrial membranes of and mitochondria isolated from Pisum sativum pea sprouts [9].

Germination of the pea seeds was performed as follows: in the control group, the seeds were rinsed with water with soap and a 0.01% solution of $KMnO_4$ and were soaked in water for 60 min; in the experimental group, the seeds were soaked in a 10^{-7} % solution of melaphen for 30 min and then in water for 30 min. On the next day, half of the seeds from the control group and half of the melaphen-treated seeds were carried over on a dry filter paper in open cuvettes. After two days of *drought,* the seeds were placed in closed cuvettes with the filter paper being moistened periodically; the seeds remained there for 7 days. On the seventh day, we counted the number of germinated seeds, measured the hypocotyl length, and isolated mitochondria.

The structural characteristics of mitochondrial membranes of sugar beet roots were investigated by the method of EPR-spectroscopy using paramagnetic spin probes: 2.2.6.6. tetramethyl-4-capriloyloxypiperidine-1-oxyl (Probe 1) and 5.6-benzo-2.2.6.6.-tetramethyl-1.2.3.4-tetrahydro-γ-carbonyl-3-oxyl (probe 2) [10]. Melaphen at concentrations of 4×10^{-12} – 2×10^{-7} M was added to the membrane suspensions (3–5 mg protein/ml) and incubated at 4°C for 30 min. A probe was put into the sample at the final concentration 10^{-5} M and incubated with mitochondria at 4°C for 30 min. The membrane samples in the absence of melaphen were used as the control. The spectra were recorded on an ER-200D SRC EPR-spectrometer (Bruker) at room temperature. From the spectra obtained, we calculated the time of rotational correlation (τ_c 10^{-10} s) of the probes by the formulas for rapidly rotating probes, that have the meaning of the period of radical reorientation for the angle $\pi/2$ [11]. The results were expressed in arbitrary units.

The respiration rates of rat liver and beet root mitochondria were recorded with the aid of the Clark-type oxygen electrode on an LP-7 polarograph. The maximum rates of oxidation of NAD-dependent substrates were determined in the incubation medium of the following composition: 0.4 M saccharose, 20 mM HEPES-Tris buffer, 5 mM KH_2PO_4, 2 mM $MgCl_2$, 5 mM EDTA, 10 mM glutamate, and 10 mM malate; pH 7.4.

In a study on the electron transfer rate at the end of the mitochondrial respiratory chain of pea sprouts, the incubation medium contained: 0,4 M saccharose, 20 mM HEPES-Tris buffer, 5 mM KH_2PO_4, 2 mM $MgCl_2$, 5 mM EDTA, 10 mM ascorbate, 60 μM potenone, 5 μM antimycin A, and 0.5 μM FCCP; pH 7.4. The antiradical activity of the compound was assessed by the method of chemiluminescence (CL). The stationary concentration of radicals during oxidation of ethylbenzene (60°) was maintained by an azo-bis-isobutylnitrile (AIBN) initiator. The CL intensity was enhanced by 9.10-dibromanthracene. The effective constant k_7 was calculated from a series of CL curves at a varied concentration of melaphen [12,13]. The results obtained were compared with data on the known antioxidants – α-tocopherol and chromane C1 (α-tocopherol analog).

Results and Discussion

As is seen from tables 1 and 2, low moisture leads to decreasing the maximum rates of oxidation of NAD-dependent substrates by pea sprout mitochondria. The rate of oxidation of glutamate+malate in the presence of FCCP (carbonylcyanide-p-trifluorometoxyphenylhydrazone) decreases from 105.0±2.1 to 75.0±3.4 natoms of O_2/mg.mitochondrial protein/min. Hence, the respiration rate with reference to Chance decreases from 2.27±0.1 to 1.7±0.2. The pretreatment of seeds with melaphen prevents from *drought*-induced change in the efficiency of oxidative phosphorylation. In addition, the preatreatment results in reduction of the rates of oxidation of NAD-dependent substrates in the presence of ADF or FCCP to the control values. The decrease in the maximum rates of oxidation of NAD-dependent substrates under low-moisture conditions may be caused by a decrease in the electron transfer rate at the end of the respiratory chain. As is seen from table 3, the rates of oxidation of ascorbate in the presence of tetramethylphenylenediamine (TMFD) by mitochondria isolated from *drought*-conditioned pea sprouts are less than the control values almost by 40%. In this case, the pretreatment with melaphen prevents from change at this site of the respiratory chain. It is safe to suppose that shifts in the physiological parameters are caused by changes in the metabolic activity of mitochondria: the pretreatment of pea seeds with melaphen stimulates the growth of sprouts (by 18-24%) both for the control and for plants growing in *drought*; however, germination of treated and untreated seeds differs significantly. Under *drought* conditions, the germination of seeds in control decreases by 46%; that of melaphen-treated seeds remains almost unchanged. Moreover, melaphen stimulates the growth of sprout roots under *drought* conditions, which is of great importance for adaptation. The adaptogenous effect of the compound may be attributed to activation of the electron transfer at the end of the mitochondrial respiratory chain, which results in inhibiting the formation of active oxygen species (AOS) at the site of complexes I and III [14,15]; the effect may account for the antistress action of the compound. Thus, melaphen treatment may cause changes in the structural properties of membranes. Indeed, the incubation of plant mitochondria with 4 x 10^{-12} M melaphen results in reduction of the lipid bilayer microviscosity by 10% (figure 1), which is evidence for the antistress effect of the compound.

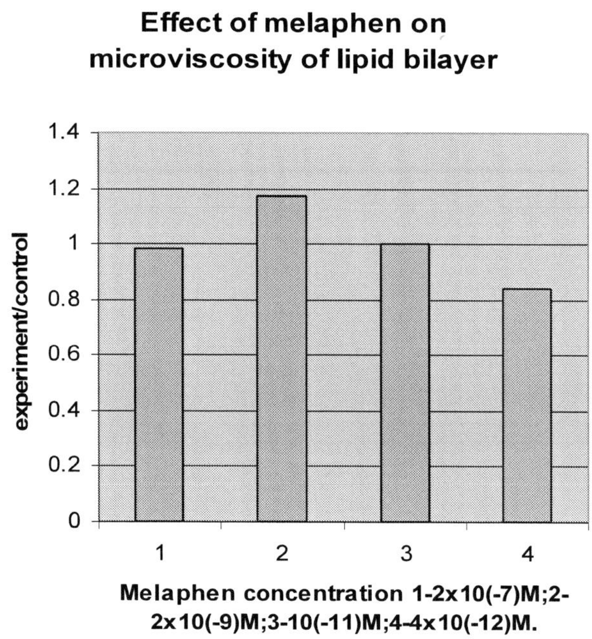

Figure 1.

Table 1. Effect of low moisture content and melaphen on maximum rates of oxidation of NAD-dependent substrates. (Respiration rates are given in natoms O_2/mg mitochondrial protein.min)

Groups	V0	FCCP
Control	19.0±2.0 (8)	105.0±2.1 (8)
Low moisture	14.0±1.5 (8)	75.0±3.0 (8)
Low moisture +melaphen	20.5±1.9 (10)	103.0±3.4 (10)

Standard incubation medium (see the method) + 10 mN glutamate, 10 mM malate, and 0.5 μM FCCP.
Designations: V_0 is the initial rate of oxidation of substrate (natom O_2/(min mg protein), FCCP – carbonylcyanide-p-trifluorometoxyphenylhydrazone.

Table 2. Effect of low moisture content and melaphen on rates of o xidation of glutamate+malate. (Respiration rates are given in natoms O2/mg mitochondrial protein.min)

Group	V0	V3	V4	V3/V4	FCCP
Control	20.0±1.5 (10)	68.0±4.1 (10)	30.0±2.0 (10)	2.27±0.1 (10)	70.0±4.6 (10)
Low moisture	12.0±2.0 (10)	48.6±3.0 (10)	40.2±1.0 (10)	1.7±0.2 (10)	48.9±3.2 (10)
Low moisture +melaphen	19.8±3.0 (10)	66.0±2.4 (10)	27.5±1.3 (10)	2.4 ± 02 (10)	75.3± 5.2 (10)

Standard incubation medium (see the method) + malate+glutamate (10 mN), ADF (200 μM), and FCCP (0.5 μM).

Designations: V_3 is the rate of oxidation of substrates (natom O_2/(min mg protein) in the presence of ADF (state 3), V_4 is the rate of oxidation of substrate after consumption of ADF (state 4).

Table 3. Effect of melaphen treatment on the rate of oxidation of ascorbate in the presence of TMFD by mitochondria isolated from pea sprouts. (Respiration rates are given in natoms O2/mg mitochondrial protein.min)

Group	V0	TMFD 200 μM	TMFD 200 μM	TMFD 400 μM
Standard conditions of germination	8.0±0.3 (5)	250.0±10 (5)	336.0±11.0 (5)	500.0±32 (5)
Low moisture	5.0±0.2 (8)	175.0±25 (8)	240.0±21 (8)	356.0±32 (8)
Melaphen treatment+Low moisture	7.0±0.4 (6)	242.0±15 (6)	346.0±17 (6)	498.0±36 (6)

The incubation medium contained: 0.4 M saccharose, 20 mM HEPES-Tris buffer, 5 mM KH_2PO_4, 2 mM MgCl2, 5мM EDTA, 10 мM ascorbate, 60 μM rotenone, 5 μM antimycin A, and 0,5 μM FCCP; pH 7.4.

In addition, melaphen exhibits antiradical properties. The effective rate constant k_7 of interaction of melaphen with peroxide radicals was found very high: $1.64x10^6$ $(Mc)^{-1}$, which is comparable with k_7 for α-tocopherol. Due to its antiradical properties, melaphen may cause changes in the structural parameters of membranes. The incubation of plant (beet root) mitochondria in the presence of $4x10^{-12}$ M melaphen results in a 10% decrease in the lipid bilayer microviscosity.

The data obtained show that a high physiological activity of the compound is associated with its effect on the physicochemical state of biological membranes by changing the lipid–protein interaction that affects the activity of membrane-associated enzymes. Due to its high antiradical activity and increasing the electron transfer rate at the final cytochromoxidase site of the mitochondrial respiratory chain, melaphen is an effective antistress agent in cases of environment-induced lipid peroxide oxidation processes.

References

[1] L.I. Chalova, O.L. Ozeretskaya : «Bioochemistry of immunity, rest and aging of plants» (I.V.Berezin).*Science,* M.., 1984. p. 41–57. (In Russian).

[2] V.V. Korol., I.G. Kirillova, T.I.Puzina.*: "Messenger Of Bashkir University"* . 2001. № 2. p. 84–86. (In Russian).

[3] Y. Morohashi: *J. Plant Physiol.* 1984, 116. p. 235–240.

[4] J.A Bunce: *Plant Cell Enviroment.* 1990, 13. p. 295–298.

[5] J.C Thomas., H.J Bohnert: *Plant. Physiol.* 1993, 103.p. 1299–1304

[6] S.G.Fattahov, V.S. Reznic, A.I. Konovalov: *«13 International conference of chemistry of phosphorus compound»*, S. .Peterburg, 2002, p. 80.(In Russian).

[7] V.I. Kostin, O.V. Kostin, V.A. Isaichev: «Condition of investigation and prospects of Melaphen application in agriculture and biotechnology», Kazan, 2006, p. 27-37 .(In Russian)

[8] S.G.Fattahov, N.L. Loseva, A.I. Konovalov, V.S. *Reznic and other* :RAS. 2004. 394. p. 127-129 (In Russian)

[9] V.N. Poov, E.K. Ruuge, A.A. Starkov. Biochemistry. 2006, 68, (7), p.910-915 (In Russian).

[10] A.N. Goloshyapov. , E.B. Burlakova *Biophisica.* 1975, 20, p. 816-821.(In Russia)

[11] A.N. KUZNETSOV. *"Method* of spin sound". M.,*Science,* 1976. 210p.(In Russian).

[12] N.V. YUDINA, S.I. Pisareva and other.*: "Chemistry of plant resources"* 1998, (4), p. 29-32.(In Russian).

[13] L.M. Pisarenko, O.T Kasaikina. *Report of RAS. Chemical series* . 2002, (3), p. 419-425.(In Russian).

[14] E Cadenas, A.Boveris, C.L Ragan., A.D.M Stoppani:. *Archiv. Biochem. Biophys.,* 1977,180, p.248- 257.

[15] V.Borutaite, G.C. Brown : *Biochemica et Biophysica Acta.* 2006, 1757, p.562-566.

In: Research Progress in Biotechnology
Editor: G. E. Zaikov, pp. 131-136

ISBN: 978-1-60456-000-8

Chapter XV

Biosensor System for Express Characterization of Strains-Destructors of Xenobiotics and Detection of Caprolactam in Aqueous Media

***I.V. Rossinskaya*[*1], *T.Z. Esikova*[2], *A.N. Reshetilov*[*2] *and O.N. Ponamoreva*[1]**

[1] Tula State University, Tula, Russia

[2] Institute of Biochemistry and Physiology of Microorganisms, Russian Academy of Sciences, Moscow Region

Abstract

Biosensor models for caprolactam detection have been developed on the basis of strains-destructors Pseudomonas putida BS394(pBS268) and Pseudomonas chlororaphis PCL 1391(pBS268). The basic analytical characteristics have been obtained for a biosensor based on Pseudomonas putida BS394(pBS268). The activity values of enzyme systems of selected stains were compared with the parameters of calibration dependences of biosensors based on these strains.

Keywords: biosensor, caprolactam, Pseudomonas, plasmids, biodegradation.

Large-tonnage productions of caprolactam and polymer materials on its basis, due to imperfection of the technological process, are associated with discharge into the environment of wastes, the main components of which are the monomer and its transformation products

* I.V. Rossinskaya: 92 Lenin pr. Tula 300600 Russia, tel. +7(4872)351840, fax +7(4872)331305, e-mail: chem@tsu.tula.ru

* A.N. Reshetilov: Pushchino, 5 Nauki pr., Moscow Region, 142292 Russia, tel. +7(4967)732788; e-mail: anatol@ibpm.pushchino.ru

[1]. In this connection, it is necessary to pay attention to the development of express monitoring for detection of hazardous levels of contamination, and studying of xenobiotic biodegradation mechanism for construct an effective method of biological purification of waste water. The goal of this work was to create a biosensor for express analysis of ε-caprolactam content in aqueous media and estimation of catabolic activity of strains-destructors of caprolactam used at formation of receptor elements.

Materials and Methods

Bacterial Strains Used in the Work

Strains *Pseudomonas putida* BS394(pBS268) and *Pseudomonas chlororaphis* PCL 1391(pBS268); plasmid-free strain *Pseudomonas putida* BS394 (Collection of the Laboratory of Plasmid Biology, Institute of Biochemistry and Physiology of Microorganisms RAS, Pushchino).

Formation of Receptor Element

Bacterial cells were precipitated by centrifugation (12,000 rpm, 10 min), washed with 30 mM sodium-potassium phosphate buffer, pH 7.6 (salts Na_2HPO_4 and KH_2PO_4 provided by Dia-M, Russia); wet cell biomass was diluted 6 times (by weight) with the buffer; the obtained suspension was mixed in the ratio of 1:1 with 2% agar gel (agar-agar bacteriological, Dia-M, Russia) cooled to 50°C. The resulting mixture was applied to a Whatman GF/A fiberglass filter (Sigma, Germany). The element (3×3 mm^2) was dried in the air for 15 min. Prepared bioreceptor element was placed onto the surface of oxygen electrode (Kronas Company, Russia) and fixed with a caprone net.

Biosensor Measurement

Measurements were made using IPC2L polarographic system (Kronas, Russia). The measuring cell was filled with 4 ml of sodium-potassium phosphate buffer, pH 7.6, salt concentration of 30 mM. Stirring was made on a magnet stirrer, 200 rpm. The measured parameter was the maximal rate of change of output biosensor signal at addition of substrates.

Determination of Specific Activity of Enzyme System for Caprolactam Degradation *In Vivo*

Cell suspension (concentration of 190 mg/ml) in buffer solution (60 mM sodium-potassium buffer, pH 7.6) was poured into a flask followed by addition of ε-caprolactam (Fluka) to a final concentration of 2.5 mM. Measurements were made at 20°C under continuous stirring. At the initial stage and in 60 min from the moment of addition, 300 mkl of solution were taken and centrifuged for 5 min at 12,000 rpm. Caprolactam concentration was determined in supernatant by thin-layer chromatography according to [2]. In parallel, caprolactam concentration was determined by the developed biosensor. The change of substrate concentration per time unit (min), as a ratio to wet cell biomass (mg), was taken as specific activity of the enzyme system *in vivo*.

Results and Discussion

Biodegradation plasmid pBS268 belongs to incompatibility group P-9 and determines caprolactam catabolism to succinate [3]. Receptor elements of the biosensor were formed using strains-destructors of caprolactam *Pseudomonas putida* BS394(pBS268) and *Pseudomonas chlororaphis* PCL 1391(pBS268). Bioreceptor elements contained equal masses of microbial cells, which provided the quantitative approach to obtaining the comparative estimates of oxidative activity of bacterial strains. Calibration curves were obtained for biosensors based on each strain (figure 1).

Parameters of calibration curves of biosensors are given in table 1.

Table 1. Parameters of calibration curves of biosensors based on *Pseudomonas putida* BS394(pBS268) and *Pseudomonas chlororaphis* PCL 1391(pBS268)

Strain	Coefficient of correlation R^2	Apparent Michaelis constant K_M, mM	Maximum rate, v_{max}, nA/s	Coefficient of sensitivity b, nA/(s*mM)
P. putida BS394 (pBS268)	0,9984	0,12±0,06	0,014±0,007	0,080±0,009
P. chlororaphis PCL1391 (pBS268)	0,9877	0,71±0,06	0,016±0,010	0,003±0,002

Bacterial strain *Pseudomonas putida* BS394(pBS268) was selected as a biosensor receptor element for caprolactam analysis in aqueous media: detection of caprolactam by this sensor was characterized by higher sensitivity. The basic analytical characteristics have been obtained for a biosensor based on *Pseudomonas putida* BS394(pBS268):

- single measurement time is about 15 min;
- limit of detection is 0.001-0.002mM, so determination of caprolactam level below its MPC (0.01mM) is possible;
- long-term sensor stability is over 15 days (the decline of microbial activity during this period is 25%);
- Operational stability of the biosensor based on *Pseudomonas putida* BS394(pBS268): convergence of results was 9.8%.

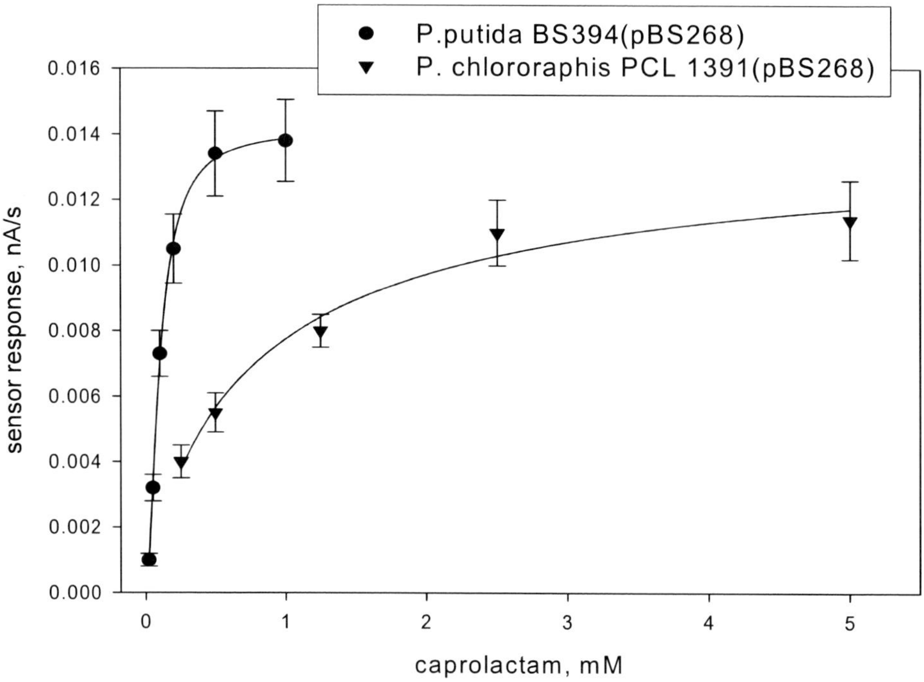

Figure 1. Calibration curves of biosensors based on *Pseudomonas putida* BS394(pBS268) and *Pseudomonas chlororaphis* PCL 1391(pBS268).

The biosensor selectivity was estimated by substrate specificity of strain *Pseudomonas putida* BS394(pBS268) (figure 2), showing that caprolactam was oxidized by these microorganisms twice more efficiently than intermediate products of its biodegradation (6-aminohexanoate, adipate [4]). This allows selective detection of caprolactam concentration in wastewater even after its biochemical treatment.

Plasmid-free strain *Pseudomonas putida* BS394 in the same conditions showed no respiratory activity toward caprolactam and its intermediates (figure 2), which makes it possible to use a differential scheme of measuring caprolactam content in aqueous media: the analysis of a sample by sensors based on plasmid and plasmid-free strains followed by subtraction of signals, which allows leveling the signals determined by the presence of interfering compounds [5].

The biosensor based on *Pseudomonas putida* BS394(pBS268), along with the method of thin-layer chromatography, was used in the work to detect caprolactam concentration to estimate the degree of its biodegradation in model systems.

The efficiency of functioning of the enzyme system catalyzing caprolactam biodegradation was assessed by the loss of substrate in a flask with cell suspension during one hour. The activities of enzyme systems of bacterial cells are given in table 2. The enzyme activity of strain *Pseudomonas putida* BS394(pBS268) is much higher than that of strain *Pseudomonas chlororaphis* PCL1391(pBS268), which is in agreement with the parameters of calibration dependences of biosensors based on these strains: the effective Michaelis constant is 6 time lower for the receptor element on the basis of *Pseudomonas putida* BS394(pBS268).

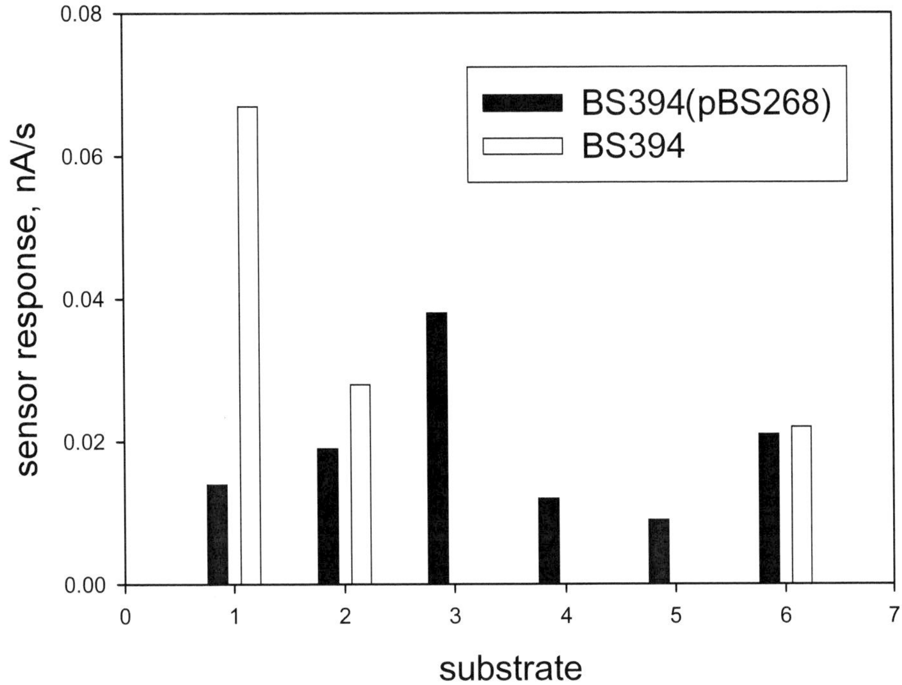

Figure 2. Substrate specificity of strain *Pseudomonas putida* BS394 (pBS268) and plasmid-free strain *Pseudomonas putida* BS394. Substrates: 1-glucose, 2-ethanol, 3-ε-caprolactam, 4 – 6-aminohexanoate, 5- adipate, 6-succinate. Substrate consentration was 1 mM.

Table 2. Specific catabolic activity of enzyme systems toward caprolactam

Strain	Specific catabolic activity of enzyme systems toward caprolactam α, μM/(mg*min)
Pseudomonas putida BS394(pBS268)	0,17±0,02
Pseudomonas chlororaphis PCL1391(pBS268)	0,06±0,03

Conclusion

Biosensor models for caprolactam detection have been developed on the basis of strains-destructors *Pseudomonas putida BS394(pBS268)* and *Pseudomonas chlororaphis PCL 1391(pBS268)*. Bacterial strain P.p. was selected as a receptor element providing optimal parameters of caprolactam detection in aqueous medium. It has been shown that caprolactam biosensor detection is characterized by high sensitivity and expressivity.

The specific activities of enzyme systems of caprolactam degradation from strains *Pseudomonas putida* BS394(pBS268) and *Pseudomonas chlororaphis* PCL 1391(pBS268) were determined and shown to coincide with the parameters of calibration dependences of biosensors based on these strains.

The work was supported by grant 2.1.1.7789 of Analytical departmental program «Development of scientific potential of the higher school (2006-2008)».

References

[1] M. Obst, A.Steinbuchel: Microbial Degradation of Poly(amino acid)s. *Biomacromolecules,* 5, 1166 (2004).

[2] Patent №4673772/25. Method of quantitative determination of caprolactam in water solution. 1991.

[3] R.P. Naumova, T.Z. Esikova, O.N. Iljinskaya, V.G. Grishchenkov, A.M. Boronin: *Microbiologia,* 57(3) (1988).

[4] Boronin A.M., Naumova R.P., Grishchencov V.G., Ilijunskaya O.N.: *FEMS Microbiol. Lett.* 22, 167 (1984).

[5] A.N. Reshetilov: Applied Biochemistry and Microbiology, 41 (5), 442 (2005).

In: Research Progress in Biotechnology
Editor: G. E. Zaikov, pp. 137-145

ISBN: 978-1-60456-000-8

Chapter XVI

Some Prospects of Utilization of Bioactive Compounds of Higher Basidiomycetes in Food Industry

M. Shamtsyan* [1], A. Panchenko[2], V. Spiridonova[1], T. Dmitriyeva[1], A. Popov[1], P.Tsvetkov[1], O. Chistova[1], Y. Tozik[1], N. Petrischev[2] and N. Denisova[3]

[1] St. Petersburg State Technological Institute (Technical University), Moskovsky prospect, 26, St. Petersburg, Russia

[2] I.P.Pavlov State Medical University, 6/8 L.Tolstoy, 197022, St. Petersburg, Russia.

[3] V.L. Komarov Botanical Institute, Professora Popova, 2, St. Petersburg, 197376, Russia

Abstract

Fruit bodies, micelia and aqueous extracts from fruit bodies and mycelia of various Higher Basidiomycetes were studied in search for reliable biological effects. *In vitro* and *in vivo* experiments were conducted. The results showed that the aqueous extracts demonstrated various types of marked biological actions: an increase generation of reactive oxygen forms by neutrophil cells of human peripheral blood; a significant mitogenic activity in a wide range of concentrations; stimulation on production of inflammatory cytokines interleukine 1-β and interleukine – 8 by peripheral blood cells; increase in the levels of anti-body forming cells, anti-tumor activity against melanoma B-16 and Erlich's ascit carcinoma. Administration of fruit bodies and mycelium of several basidiomycetes accompanied with hyperlipidemic diet caused lowering of cholesterol levels in blood and liver and decrease of the values of atherogenity coefficient. Aqueous extracts of submerge mycelia possessed significant antioxidant and SOD-like activity. Milk clotting enzymes from higher basidiomycetes with high clotting and law general proteolytic activities were characterised.

* M. Shamtsyan: e-mail: shamtsyan@yahoo.com

Keywords: Higher Basidiomycetes, immunomodulation, anti-tumor activity, cholesterol-lowering effect, antioxidant activity, milk-clotting effect.

Introduction

From many-secular experience of the folk medicine in various countries it is well known that in addition to their nutritional value, higher Basidiomycetes have gained a wide spectrum of therapeutic and prophylactic usage. For thousands of years humans have used medicinal mushrooms for their healing potential. Even shamans of Paleolithic peoples knew and used mushrooms as powerful medicines to fight illnesses.

Greeks were concerned to mushrooms as a source of strength for warriors before the battle. Pharaohs were appreciating mushrooms as delicacies. According to the Hieroglyphics of 4600 years ago ancient Egyptians believed that mushrooms are the plant of immortality. The delicious flavor of mushrooms intrigued the pharaohs of Egypt so much that they decreed that mushrooms were food for royalty and that no commoner could ever touch them. This assured themselves the entire supply of mushrooms. Romans were considering them as the meal of gods, and were serving them only for great fists. Vikings were going to berserk tasting them, while the people of South America were overcoming the barrier between the body and mind by their magic hallucinogen action.

In various civilizations throughout the world, including Russia, China, Greece, Mexico and Latin America, mushroom rituals were practiced. Many believed that mushrooms had properties that could produce super-human strength, help in finding lost objects and lead the soul to the realm of the gods.

For the centuries mushrooms were used in different cultures for their medical and tonic properties. The major part of the traditional knowledge of mushrooms medical properties is coming from the Far East. The richest traditions of use of edible mushrooms for healing purposes still have China and Japan. Naturally, during the evolutional process the community step by step accumulated and assimilated the traditions of the folk medicine.

Western countries are just starting to use the variability and tremendous potential of medical mushrooms.

Historically, mushrooms were gathered from the wild for consumption and for medicinal use. China has been the source of many early cultivations of mushrooms, e.g. *Auricularia auricula* (600 AD), *Flammulina velutipes* (800 AD), *Lentinus edodes* (1000AD) and *Tremella fuciformis* (1800). *Agaricus bisporus* was first cultivated in France in *c* 1600 while *Pleurotus ostreatus* was first grown in US in 1900. While mushroom cultivation now spans many centuries, it is only over the last 2-3 decades that there have been major expansions in basic research and practical knowledge leading to the creation of a major worldwide industry.

In the second half of 20th century technologies of mushroom cultivation were strongly developed, and at the end of the century the volume of world mushroom production was estimated as $ 18 billion, which is almost equal to the world production of coffee (Chang, 1999).

At present there are at least 270 species of mushroom that are known to have various therapeutic properties. On the world pharmaceutical market today exists a large arsenal of mushroom compounds, which are representing a new class – so called "nutriceutics" or "functional food additives". The researches for new natural substances, able to enhance "biological functioning of human organism", and so, to enhance the quality of our life are actively going on. In the recent years medics and dietologists are more and more often referring to food products as a potential source of such substances.

While much attention has been drawn to various immunological and anti-cancer properties of mushrooms (Jong and Donovick, 1989, Mizuno, 1995), they also offer other potentially important therapeutic properties including antioxidants, anti-hypertensive, cholesterol-lowering, liver protection, fibrinolytic, anti-inflammatory, anti-diabetic, anti-viral, anti-microbial and others (Hobbs, 1995, 1998, Mizuno et al., 1999, Gunde-Cimerman, 1999, 1999, Mau et al., 2002 and the others).

Objects of our studies were fruit bodies mycelium and cultural filtrate of submerge cultivation of species of mushroom species *Armillaria mellea, Auricularia auricula-judae, Bjerkandera adusta, Coprinus sp., Inonotus obliquus, I. radiatus, Flammulina velutipes, Fomes fomentarius, Fomitopsis pinicola, Grifola frondosa Ganoderma lipsiense, G. Lucidum, Lentinus edodes, Piptoporus betulinus, Pleurotus cornucopiae, P. ostreatus, Trametes suaveolens, Tremella foliacea, Tricholoma portentosum*, and also aqueous extracts from cultural mycelium and fruit bodies.

Fruit bodies were collected in the forests of St. Petersburg region of Russia, or purchased from the local market, and several species of basidiomycetes were introduced into culture. The mycelium was obtained by a submerged cultivation of strains on mash (4°B) at 28°C. For immunological assays mycelium and fruit bodies were treated using the method described by Wang et al. (1993) with several modifications. The biomass was dried at 60°C, crumbled up and treated with 85% ethanol at 80°C during 8 hours. The sediment was extracted twice by boiling water for 3 hours. The obtained filtrate was evaporated under vacuum, and the polysaccharide fractions were precipitated by 5 volumes of ethanol, dialyzed against distilled water and freeze dried. Obtained powder extracts, before tested on immunomodulating and anti-tumor activities, were stored at the temperature of 18 – 20°C.

To study hypocholesterolic action fruit bodies and mycelia of basidiomycetes were dried at 50°C and crumbled up into powder.

For the studies of antioxidant activities submerged mycelium was dried at 50°C, crumbled up and extracted with distilled water at room temperature, for 5 hours.

For the assays of milk-clotting activity cultural filtrates of submerge cultivation of basidiomycetes were used.

Immunomodulating and Anti-Tumor Activity

The medicine of passed centuries side by side with the real general progress leaved to the man a round of serious problems. One of such problems remains a problem of oncology diseases.

Numerous investigations of anti-tumor activity for various natural compounds, including mushroom origin compounds, contributed in the search of the possible ways to solve the global problem. Medicinal mushrooms have become even more widely used as traditional medicinal ingredients for the treatment of various diseases and related health problems largely due to the increased ability to produce the mushrooms by artificial methods. As a result of large numbers of scientific studies on medicinal mushrooms especially in Japan, China and Korea, over the past three decades, many of the traditional uses have been confirmed and new applications developed. Within half a century mycology has made a profound impact, from the first articles on research of anti-tumor activity of extracts from carpophores of various species of macromycetes (Kanatsu et al., 1963, Gregory et al., 1966; Espenshade, Griffith, 1966) to creating "basidial" remedies and introducing them to the world pharmaceutical market (Mizuno, (ed.), 1995, Mizuno, 1999, Wasser and Weis, 1999, and many other).

For the past decades, the strategy of research on the anti-tumor and immunomodulating properties in high basidiomycetes logically headed on a general course of complex research of therapeutic, prophylactic, nutritious and other benefits of macromycetes

The list of such preparations was opened with the famous species *Schizophyllum commune*, *Trametes versicolor* and "glorious" in the East species *Ganoderma lucidum*, *Lentinus edodes*, *Grifola frondosa* and other.

Last decades the strategy of investigations of the antitumour features in high basidiomycetes logically disembogued in the general course of complex research of therapeutic, prophylactic, nutritious and other goodness of macromycetes. In the first place it had respect to the most popular in mushroom industry species from genus *Agaricus, Pleurotus, Lentinus, Ganoderma, Grifola, Flammulina, Volvariella, Auricularia, Tremella* (S.-T.Chang, 1999). However, many other species warrant further research.

The main antitumour compounds presently isolated from mushroom fruit-bodies, submerged cultural mycelial biomass or liquid culture broth have been identified as either water soluble β-D-glucans with heterosaccharide chains of xylose, mannose, galactose or uronic acid or β-D-glucan-protein complexes – proteoglycans.

In general, there is normally a higher level and number of different polysaccharides extracted from fruit-bodies than from the micellia or culture broth. Nevertheless, important direction for future studies on mushroom polysaccharides will be by submerged fermenter culture to produce reliable, consistent and safe products.

We conducted an experimental study on immunomodulating and anti-tumor activities of fruit bodies and submerged mycelia of several mushrooms: *Armillaria mellea* (Vahl: Fr.) P. Kumm., *Bjerkandera adusta* (Willd.: Fr.) P. Karst., *Flammulina velutipes* (Curt.: Fr.) Singer, *Fomes fomentarius* (L.: Fr.) Fr., *Fomitopsis pinicola* (Sw.: Fr.) P. Karst., *Ganoderma lipsiense* (Batsch) G.F. Atk. (= G. applanatum (Wallr.) Pat.), *Piptoporus betulinus* (Bull.: Fr.) P. Karst., *Pleurotus cornucopiae* (Pau. ex Pers.) Rollan (= P. citrinopileatus Singer), *P. ostreatus* (Jacq.: Fr.) P. Kumm., *Trametes suaveolens* (Fr.) Fr., *Auricularia auricula-judae* (Bull. ex St. Amans) Wettst., *Inonotus obliquus* (Pers.: Fr.) Piliÿt, *Inonotus radiatus* (Sowerby: Fr.) P. Karst., *Tremella foliacea* Pers.: Fr., *Tricholoma portentosum* (Fr.) Qui¦l., *Grifola frondosa* (Dicks.: Fr.) Gray, *Lentinus edodes* (Berk.) Singer.

Some of the species tested in our work were never studied before for immunomodulating activity.

The content of total saccharides in mushroom aqueous extracts was analyzed and was estimated to be in a range of 62 – 80 % of dry matter, for all of the specimens. The protein content in the extracts was of up to 30 %.

It was determined, that mushroom extracts have expressed immunostimulating action, involving in response almost of all immunocompetent cells. We also marked mitogenic effect of mushroom extracts on lymphoid cells, co-stimulating action on monocytic cells of a peripheral blood synthesizing pro-inflammatory cytokines, and also ability of these preparations to promote main cells of inflammatory response - neutrophils, causing not only chemotaxis but also oxygen burst showing an activation of phagocytic response. At the same time it is clear, that the extracts of diferent mushrooms can stimulate different immunocompetent cells. Uptake of aqueous extracts of mycelia of various mushrooms demonstrates the doze-depending increase of the levels of antibody forming cells in the experimental groups of mice.

All of the data obtained during our investigations suggest that aqueous extracts of studied basidiomycetes have marked biological effects. The highest activity was determined for the extracts from *Bjerkandera adusta*, *Pleurotus cornucopiae*, *Pleurotus ostreatus*, *Auricularia auricula-judae*, *Inonutus radiatus*, *Tricholoma portentosum*, *Grifola frondosa* and *Lentinus edodes*. The levels of stimulation of different immunocompetent cells, significantly varies for the extracts of different mushroom species.

The outcomes obtained in the experiments *in vitro* have predetermined check of these extracts on antitumor effect in experiment *in vivo*. As a model of tumor growth two of the most malignant tumors: melanoma B16 and Ehrlich's ascit carcinoma were selected.

The experiments have been done on16 C57BL female mice (average body weight 23 g) purchased from animal facility "Rappolovo", Russian Academy of Science. Animal were divided into control and experimental groups. The control group was given usual fodder. The experimental group animals were given modified fodder with extract. During two week periods from a beginning of extract administration the average body weight did not differ significantly as compared to control.

At 15 day melanoma B16 was transplanted to all animals (10^6 cells of a melanoma B16 subcutaneously to the right abdomen side). In further measurement of parameters of tumor growth and life duration was conducted.

The progressive and irregular decrease of average body weight was noted in both groups after tumor transplantation but less in-group received extract. At 26 day after tumor transplantation average body weight was significantly higher in experimental group. This is the evidence of decrease the body intoxication by mushroom supplements, and thus the increase of the quality of life. Among tested species most effective were supplements from *P. ostreatus*. In the control group, average tumor size increased during tumor development. In experimental group average tumor size decrease 30-80 % was marked with maximal effect during first 14 days. The average and maximal duration of life in check group reliably is lower in comparison with experimental group.

Thus, the extract from *P.ostreatus* fruit bodies makes reliable inhibiting effect on melanoma B-16 growth and decreases a manifestation of tumor intoxication as it confirms by the weight differences between control and experience mice. But it has almost no influence on the process of development of another tumor. The study of the influence of mushroom supplements on the development of Erlich's ascit carcinoma shows a significant increase in the life time of experimental mice, but the most prolongation was detected for the group were fodder was supplemented with *C. unicolor* extract.

These effects perhaps are connected with the differences in stimulation of immune system, which has a large importance for a pathogenesis of the tumor growth.

Cholesterol-Lowering Effect

It is known, that hypercholesterolemia increases the risk of heart disease. Elevated levels of circulating cholesterol cause deposits to form inside blood vessels. These deposits can result in a disease process called atherosclerosis, which can cause blood clots to form that will ultimately totally stop blood flow. If this happens in the arteries supplying the heart, a heart attack will occur. If it happens in the brain, the result is a stroke where a portion of brain tissue dies. Atherosclerosis causes more deaths from heart disease than any other single condition. The most common cause of elevated serum cholesterol is eating foods that are rich in saturated fats or contain high levels of cholesterol.

Cholesterol has been divided into two major categories: low-density lipoprotein (LDL), the so-called "bad" cholesterol, and high-density lipoprotein (HDL), the so-called "good" cholesterol.

We studied the influence of mushroom additives to hypercholesterolic diets on the levels of total cholesterol and high-density lipoproteins in rats. The obtained results are demonstrating, that the addition of fruit bodies of Pleurotus ostreatus to hyperlipidemic diets of rats caused significant decrease in a level of total cholesterol and increase of a level of useful α-cholesterol in blood of experimental animals and almost normalizing the coefficient of atherogenity.

Also it has been established, that the additive to a diet of rats of submerged mycelium of Ganoderma lucidum considerably lowered a level of the total cholesterol in blood of rats. Addition to a diet of experimental animals of submerged mycelium of Flamulina velutipes only slightly decreased the level of total cholesterol. The level of α-cholesterol by mycelium additives of both studied basidiomycetes was not affected.

Extracts also were significantly decreasing the levels of triglycerides in blood.

Studies of the levels of cholesterol in rat livers, shows that fruit bodies of Pleurotus ostreatus are efficiently inhibiting the syntheses of cholesterol in liver.

As a result of conducted studies it is possible to draw a conclusion that the additives to hyperlipidemic diets of a biomass of some mushrooms leads to essential decrease in a risk of development of atherosclerosis. The highest hypocholesterolic activity was detected for the fruit bodies of Pleurotus ostreatus and mycelium of Ganoderma lucidum.

Anti-Oxidant Activity

Antioxidants, scavenge for free radicals, and consequently are a very special group of nutritional supplements. They very special because they play such a very important role in protecting the body from the formation of very dangerous substances referred to as free radicals.

Free-radical oxidation is an important biochemical process of transformation of oxygen, lipids, nucleic acids, proteins and other compounds due to the action of free radicals. Free radicals are very unstable due to the existence of at least one unpaired electron. and react quickly with other compounds, trying to capture the needed electron to gain stability. Generally, free radicals attack the nearest stable molecule, "stealing" its electron. When the "attacked" molecule loses its electron, it becomes a free radical itself, beginning a chain reaction. Once the process is started, it can cascade, finally resulting in the disruption of a living cell.

Some free radicals arise normally during metabolism. Sometimes the body's immune system's cells purposefully create them to neutralize viruses and bacteria. However, environmental factors such as pollution, radiation, and cigarette smoke and herbicides can also spawn free radicals.

Normally, the body can handle free radicals, but if antioxidants are unavailable, or if the free-radical production becomes excessive, damage can occur. Of particular importance is that free radical damage accumulates with age.

Antioxidants are molecules, which can safely interact with free radicals and terminate the chain reaction before vital molecules are damaged. They act as scavengers, helping to prevent cell and tissue damage that could lead to cellular damage and disease.

Antioxidants can be found in a number of important diet and supplement related compounds. For example there are four enzymes that have a strong tendency to neutralize free radicals including super oxide dismutase (SOD), Methionine reductase, catalase, and Glutathione peroxydase. These four enzymes are also essential enzymes, which the body makes on its own. However, the body often is unable to, for various reasons make sufficient quantities to control effectively the ravages all of free radical damage. There are a number of other nutritional substances, which also act as antioxidants. The substances include vitamin A, beta-carotene, vitamin C, and vitamin E as well as the mineral selenium and the others.

Antioxidants are intimately involved in the prevention of cellular damage - the common pathway for cancer, aging, and a variety of diseases.

The body cannot manufacture many of antioxidants at all, or at necessary levels, so they must be supplied in the diet. Consequently, it becomes very important for many individuals to supplement their diet with compounds containing abundant levels of antioxidants. Synergistically, the body, food intake, the intake of supplements, work together to ward off the deleterious affects all of free radicals.

In our studies antioxidant activity of aqueous extracts of submerge mycelia of higher basidiomycetes was examined. All tested species demonstrated significant antioxidant activity towards primary and secondary antioxidants.

Experiments to define SOD-like activity shows, that all of the tested species, except extract from the mycelium of F. velutipes possess high SOD-like activity.

Milk-Clotting Activity

Higher basidial mushrooms can represent an interest for the food industry, not only as a source of biologically active supplements, but also as a source of wide range of useful enzymes.

Milk clotting enzymes from higher basidiomycetes are a promising source to substitute rennin in cheese making. Substitution of costly rennet enzymes by microbial proteases of specific action is economically very efficient and perspective.

The quality of cheese significantly depends from enzyme preparations, which are used for milk clotting. Mostly animal enzymes extracted from rennet are used for this purpose. Although various microbial coagulants are developed as inexpensive substitutes for them, but they does not found wide utilization in the countries of developed cheese making, were are used mainly for the production of cheddar type cheeses.

The aim of our studies was a screening of the producers of milk clotting enzymes among higher basidiomycetes. The fruit bodies, submerged mycelium, grown on various media, as well as the culture broth of several higher basidiomycetes were screened for milk clotting activity. Mushrooms, which fruit bodies, mycelium or culture broth posses high milk clotting and low proteolytic activities were selected.

During the study a basidiomycet, identified as Coprinus lagopides. which was producing in submerge culture high milk-clotting activity and low general proteolytic activity was selected. Optimization of conditions cultivation and composition of fermentation media allowed to increase yield and activity of the enzyme.

Ultrafiltration process permits considerably to increase specific activity of the enzyme preparation. As a result of action of enzyme on casein of milk the dense milk clot without bitterness was formed. Considering the received results, it is possible to assume, that after the certain purification and possible separation from existing proteolytic enzymes, obtained preparation could have prospects to fulfill the demands to high quality renal preparations.

Molecular weight of the studied milk-clotting enzyme was determined as 39±0,5 kDa. were performed to define kinetic constants. Based on the results of enzymatic reactions of milk-clotting, kinetic constants of enzyme were defined as: for Mikhaelis constant: $Km=(0{,}70\pm0{,}01)*10^{-3}$ mol/l, and for maximum specific rate of the enzymatic reaction: $Vmax= (0{,}251\pm0{,}006)*10^{-6}$ mol/(min*mg).

References

[1] Chang S.T., (1999). Global impact of edible and medicinal mushrooms on human welfare in the 21st century: nongreen revolution. *Int. J. Med. Mushr.,* 1, 1-7.

[2] Espenshade M.A., and Griffith E.W., (1966). Tumor-inhibiting basidiomycetes. Isolation and cultivation in the laboratory. *Mycologia,* 58, 511-517.

[3] Gunde-Cimerman, N. (1999). Medicinal value of the genus Pleurotus (Fr.) P. Karst. (Agaricales S.I., Basidiomycetes). *Int. J. Med. Mushr.,* 1, 69-80.

[4] Gregory F.J., Healy E.M., Agersborg H.P.K., Jr., and Warren G.H. (1966). Studies of antitumor substances produced by Basidiomycetes. *Mycologia,* 58, 80-90.

[5] Hobbs, C. (1995). Medicinal Mushrooms: An Exploration of Tradition, Healing and Culture. Botanica Press, Santa Cruz, CA. 347p.

[6] Jong, S.C. and Donovick, R. (1989). Antitumour and antiviral substances from fungi. *Advances in Applied Microbiology*. 34, 183-262.

[7] Kanatsu, N., Terakawa, H., Nakanishi, K. and Watanabe, Y. (1963). Flammulin, a basic protein of Flammulina velutipes with antitumour activities. *Journal of Antibiotics. Ser. A.*, 16, 139-143.

[8] Mau J.L., Lin H.C., Chen C.C. (2002). Antioxidant properties of several medicinal mushrooms. *Journal of Agricultural and Food Chemistry,* Vol. 50, 21, 6072 – 6077.

[9] Mizuno, T. (1995). Yamabushitake, Hericium erinaceus: bioactive substances and medicinal utilisation. *Food Review International. 11*, 173-178.

[10] Mizuno, T. (1999). Medicinal effects and utilisation of Cordyceps (Fr. Link (Ascomycetes) and Isaria Fd. (Mitosporic Fungi) Chinese caterpillar fungi "Tochukaso" (Review). *International Journal of Medicinal Mushrooms*. 1, 251-261.

[11] MizunoT. (1999). The extraction and development of antitumor-active polysaccharides from medicinal mushrooms in Japan (review). *Int. J. Med. Mushr.*, 1, 9 – 29.

[12] Wang G., Zhang J., Mizuno T., Zhuang C., Ito H., Mayuzumi H., Okamoto H. and Li J., (1993). Antitumor active polysaccharides from Chinese Mushroom Songshan Lingzhi, the fruiting body of Ganoderma tsugae Biocsi. *Biotech. Biochem.* 57, 894-900.

[13] Wasser S.P., and Weis A., (1999). Medicinal properties of substances occuring in higher basidiomycetes mushrooms: current perspectives (review). *Int. J. Med. Mushr.*, 1, 31-62

In: Research Progress in Biotechnology
Editor: G. E. Zaikov, pp. 147-157

ISBN: 978-1-60456-000-8

Chapter XVII

Study of Immunomodulating and Antitumour Activity of Extracts of Higher Basidiomycetes

***Vera Spiridonova*[1]*, Valentina Konusova*[2]*, Andrey Panchenko*[3]*, Nikolay Petrischev*[3]*, Nina Denisova*[4] *and Mark Shamtsyan**[*1]

[1] Department of Technology of Microbiological Synthesis, St.Petersburg State Institute of Technology (Technical University), 26 Moscovsky pr., 198013 Saint Petersburg, Russia;
[2] Laboratory of Immunopharmacology, State Research Institute of Extra Pure Biopreparations, 7 Pudozhskaya str., 197110 Saint Petersburg, Russia;
[3] Department of Pathological Physiology, Acad.I.P.Pavlov St.Petersburg Medicinal University, 6/8 L.Tolstoy str., 197022 Saint Petersburg, Russia;
[4] Laboratory for Biochemistry of Mushrooms, V.L.Komarov Botanical Institute, Russian Academy of Sciences, 2 Prof.Popov Str., 197376 Saint Petersburg, Russia

Abstract

Aqueous extracts from fruit bodies and mycelia of various Higher Basidiomycetes were studied in search of immunomodulating activity. In vitro and in vivo experiments were conducted. The results showed that the aqueous extracts demonstrated various types of marked biological actions: an increase generation of reactive oxygen forms by neutrophil cells of human peripheral blood; a significant mitogenic activity in a wide range of concentrations; stimulation on production of inflammatory cytokines interleukine 1-β and interleukine – 8 by peripheral blood cells; increase in the levels of anti-body forming cells, anti-tumor activity against melanoma B-16 and Ehrlikh's ascit carcinoma.

Keywords: Higher Basidiomycetes, β-glucans, immunomodulating effect, antitumor effect.

[*] Mark Shamtsyan e-mail: *shamtsyan@yahoo.com*

Introduction

For the centuries mushrooms were used in different cultures for their medical and tonic properties. The major part of the traditional knowledge of mushrooms medical properties is coming from the Far East.

Historically, mushrooms were gathered from the wild for consumption and for medicinal use. While mushroom cultivation now spans many centuries, it is only over the last 2-3 decades that there have been major expansions in basic research and practical knowledge leading to the creation of a major worldwide industry.

Bological activity of aqueous extracts from fruit bodies and submerged mycelia of various higher Basidiomycetes were studied.

Materials and Methods

Objects of our study were fruit bodies mycelium and cultural filtrate of submerge cultivation of species of mushroom species *Armillaria mellea*, *Auricularia auricula-judae*, *Bjerkandera adusta*, *Coprinus sp.*, *Inonotus obliquus*, *I. radiatus*, *Flammulina velutipes*, *Fomes fomentarius*, *Fomitopsis pinicola*, *Grifola frondosa Ganoderma lipsiense, G. Lucidum, Lentinus edodes, Piptoporus betulinus*, *Pleurotus cornucopiae*, *P. ostreatus*, *Trametes suaveolens*, *Tremella foliacea*, *Tricholoma portentosum*, and also aqueous extracts from cultural mycelium and fruit bodies.

Fruit bodies were collected in the forests of St. Petersburg region of Russia, or purchased from the local market, and several species of basidiomycetes were introduced into culture. The mycelium was obtained by a submerged cultivation of strains on mash (4°B) at 28°C. For immunological assays mycelium and fruit bodies were treated using the method described by Wang et al. (1993) with several modifications. The biomass was dried at 60°C, crumbled up and treated with 85% ethanol at 80°C during 8 hours. The sediment was extracted twice by boiling water for 3 hours. The obtained filtrate was evaporated under vacuum, and the polysaccharide fractions were precipitated by 5 volumes of ethanol, dialyzed against distilled water and freeze dried. Obtained powder extracts, before tested on immunomodulating and anti-tumor activities, were stored at the temperature of 18 – 20°C.

Immunological Reagents

For immunological studies, Hank's salt solution, luminol, phorbolmiristate acetate, agarose EEO, formyl-peptide fMLP 1640, cow fetal serum, prodigiozan, concanavalin A (all "Sigma") and fresh donor blood were used.

Blood Cells Isolation

To obtain mononuclear and granulocyte cells from the donor blood, stabilized with heparin solution (20 IU per 1 ml of blood) in ficoll-paque gradient, the standard technique described by Boyum (1968) was used. Blood plasma was layered on to ficoll-paque gradient and centrifuged during 40 min on 400 g. The mononuclear cells were collected from the interphase between plasma and ficoll-paque. The neutrophilic granulocytes were gathered from the precipitate after isolation of mononuclear cells by a hypotonic shock with distilled water during 30 sec.

Study of the Cytokin-Inducing Activity of the Preparations

Mononuclear cell suspension (concentration 5 10^6 ml^{-1}) resuspended in RPMI 1640 medium and included 10% fetal serum was placed in the 96-well flat-bottomed microplates ("Costar") by 0.1 ml in each hole. Simultaneously, the investigating preparations in appropriate solutions were added in the same holes. The Russian preparation of lipopolysaccharide nature prodigiosan, strong inductor of antiinflammatory cytokines was used as the positive control. The microplates were placed in CO_2-incubator during 24 h, after that the amount of cytokines in the cultural medium was detected by immunoenzymatic method (Dike, Forlin, 1998). Test-systems made by the company "Cytokine" (Saint Petersburg, Russia) were used.

Study of the Preparations Influence on the Production of Oxygen Active Forms by Phagocyting Cells

The formation of oxygen active forms by whole blood cells was detected with luminoldependent chemiluminescence method (LDCL) (Tono-Oro et al., 1990). The determination of LDCL was carried out in the 96-well white nontransparent microplates "Costar" with the device Victor-2 (Finland), connected with PC and provided sequent registration of chemiluminescence. The reaction result was termed as light sum (sum of impulse accumulated during the certain time period).

The base luminol solution (10^{-2} M) was prepared in DMSO and stored at 4°C. The working solution (10^{-4} M) was prepared fresh by diluting the base solution with Hank's solution (pH 7.2).

The experiment was carried out follow scheme: 120 μl of the Hank's solution, 20 μl of the fresh donor blood, 40 μl of the working luminol solution and 20 μl of the investigating preparation in the appropriate dilutions were placed in the each well of the 96-well microplates. Each test was prosecuted in the three parallels. The solution of phorbol-miristate acetate (PMA) in the end concentration 10 ng ml^{-1} was used as the standard positive control. The reaction was fixed during 1 h.

The Preparations Influence on the Cell Proliferation

The proliferative processes in the cells were studied with the reaction of blust-transformation of spleen cells of CBA strain mice (farm Rappolovo, Saint Petersburg, Russia). The mice were slaughtered by method of cervical dislocation, the spleen was extracted, grinded up with the scissors, washed out by syringing the cultural medium RPMI 1640, and then washed clean by twice centrifuging. The cell suspension (concentration 2 10^6 ml^{-1}) in the RPMI 1640 medium containing 10% fetal serum was prepared. The cell suspension was placed by 0.1 ml in the 96-well microplates "Costar" together with the investigating preparations. Concanavaline A (Con A) in the concentration 0.1 and 0.2 mg ml^{-1} was used as the standard inductor of proliferation. The microplates were placed in the CO_2-incubator (CO_2 5%, 37°C) during 4 days. After 3 days the labeled tritium ^3H thymidine was added in the cultural medium. Result of reaction was estimated in impulses per min with using beta-meter.

Results and Discussion

Immunomodulating and Anti-Tumor Activity

The main antitumour compounds presently isolated from mushroom fruit-bodies, submerged cultural mycelial biomass or liquid culture broth have been identified as either water soluble β-D-glucans with heterosaccharide chains of xylose, mannose, galactose or uronic acid or β-D-glucan-protein complexes – proteoglycans.

We conducted an experimental study on immunomodulating and anti-tumor activities of fruit bodies and submerged mycelia of several mushrooms: *Armillaria mellea* (Vahl: Fr.) P. Kumm., *Bjerkandera adusta* (Willd.: Fr.) P. Karst., *Flammulina velutipes* (Curt.: Fr.) Singer, *Fomes fomentarius* (L.: Fr.) Fr., *Fomitopsis pinicola* (Sw.: Fr.) P. Karst., *Ganoderma lipsiense* (Batsch) G.F. Atk. (= G. applanatum (Wallr.) Pat.), *Piptoporus betulinus* (Bull.: Fr.) P. Karst., *Pleurotus cornucopiae* (Pau. ex Pers.) Rollan (= P. citrinopileatus Singer), *P. ostreatus* (Jacq.: Fr.) P. Kumm., *Trametes suaveolens* (Fr.) Fr., *Auricularia auricula-judae* (Bull. ex St. Amans) Wettst., *Inonotus obliquus* (Pers.: Fr.) Piliÿt, *Inonotus radiatus* (Sowerby: Fr.) P. Karst., *Tremella foliacea* Pers.: Fr., *Tricholoma portentosum* (Fr.) Qui¦l., *Grifola frondosa* (Dicks.: Fr.) Gray, *Lentinus edodes* (Berk.) Singer.

The content of total saccharides in mushroom aqueous extracts was analyzed and was estimated to be in a range of 62 – 80 % of dry matter, for all of the specimens. The protein content in the extracts was of up to 30 %.

Primary evaluation of biological activity of studied mushroom aqueous extracts was done by test-analysis of the reactive oxygen species generated by cells of a peripheral human blood by a method of luminoldepended chemiluminescence. The test was selected on the basis of the supposition about stimulating effect of extracts on the system of phagocytes, since it is known, that the compounds of polysaccharide nature boost the phagocytosis. When activating the phagocytes of the blood the cascade of chain reactions is started accompanied with the increased oxygen consumption from the environment. Thus the cells pass on hexosomonophosphate path of respiration. The final products of the reaction contained the reactive oxygen species, which being combined with luminol caused a luminescence registered at a wave-length 425 nm.

The extracts from fruit bodies and / or mycelia of all mentioned species were studied in concentrations 0.001-1.0 mg ml^{-1}. Many of the experienced samples of mushrooms increased production of reactive oxygen species by neutrophils of a peripheral human blood. The maximal effect of all experienced extracts observed in a dose at 0.1 - 1.0 mg/ml. On the figure 1. data obtained for a doze of 1.0 mg/ml are presented. The effect of fruit bodies extracts was slightly higher, then the effect observed for mycelia extracts of same species. In both cases the maximal effects were observed for extracts of mushrooms of *P. ostreatus* and *P. cornucopiae.*

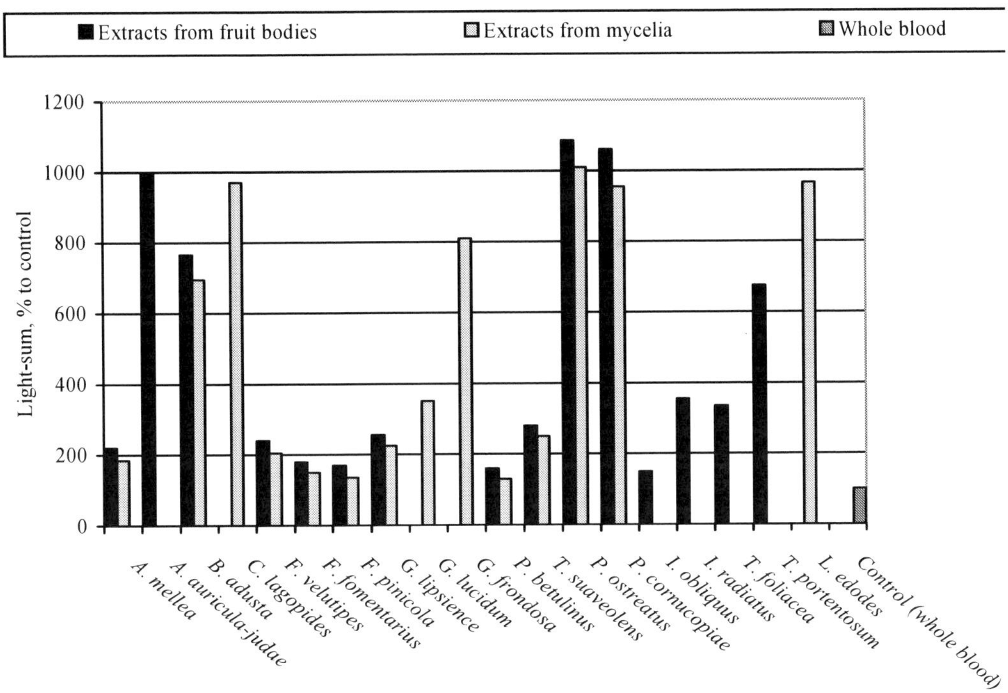

Figure 1. The effect of extracts from mushrooms fruit bodies and mycelia on generation of reactive oxygen species by human blood neutrophils.

The mitogenic activity of fruit bodies extracts for mushrooms that possessed higher activity in generation of the reactive oxygen was studied (table 1). The research was carried out in blast-transformation reaction (BTR) on spleen cells of CBA mice.

Table 1. Influence of mushroom extracts on blast-transformation reaction of leucocytes from CBA mice spleen

Extracts from fruit bodies of mushroom species	Preparation concentration, mg ml^{-1}						
	0.5	0.25	0.125	0.0625	0.03125	0.0156	0.0078
	Level of blust-transformation reaction, % of control						
Bjerkandera adusta	920	970	900	660	580	510	460
Inonotus radiatus	270	280	290	230	160	130	110
Ganoderma lipsiense	190	180	210	140	120	110	80
Tricholoma portentosum	340	360	320	270	190	140	130
Pleurotus ostreatus	640	650	640	600	540	550	380
Pleurotus cornucopiae	150	460	540	550	590	480	360

Level of spontaneous BTR (control) - 8.3 thousands imp min^{-1}
Index of stimulation of BTR as response on ConA (0.1 mg ml^{-1}) – 390% of control.

From the data of table 1 it is clear that the extracts of all tested mushrooms have marked mitogenic activity in a wide range of concentrations. In this tests the maximal activity levels were detected for the extracts of *B. adusta*. Both species of *Pleurotus* mushrooms were also active in a wide range of concentrations.

It was studied the ability of extracts from fruit bodies of several mushrooms to participate in synthesis by human blood cells of immunoregulating proteins – cytokines, particularly of interleukine-1β and interleukine- 8, being a key mediator of inflammatory and immune response of an organism. The ability of a mushroom extracts to influence lipopolysaccharide induced cytokine production was studied. The data obtained are shown in table 2.

It can be seen from the table 2 that preparations display a co-stimulating effect on both cytokines production by peripheral blood cells. Most effective were preparations *from I. radiatus* and *T. Portentosum*

The studies of the influence of uptake of aqueous extracts of mycelia of B. adusta, P. ostreatus and P. cornucopiae demonstrate the doze-depending increase of the levels of antibody forming cells in the experimental groups of mice (figure 2.)

The outcomes obtained in the experiments *in vitro* have predetermined check of these extracts on antitumor effect in experiment *in vivo*. As a model of tumor growth melanoma B16 and Ehrlich's ascit carcinoma were selected.

Table 2. Influence of an extracts of mushroom fruit bodies on production of inflamatory cytokines by human blood cells

Preparation concentration, mg ml^{-1}	Exreacts from fruit bodies									
	B.adusta		I. radiatus		P. cornucopiae		P. ostreatus		T. portentosum	
	Level of cytokine production, % of control									
	IL 1-β	IL-8	IL 1-β	IL-8	IL 1-β	IL-8	IL 1-β	IL-8	IL 1-β	IL-8
5.0	320	310	370	>500	210	220	330	150	280	360
0.5	210	180	270	>500	280	310	280	140	330	>500
0.05	140	130	140	420	230	180	190	130	260	170
0.005	106	103	120	190	150	110	150	110	190	104
0.0005	100	100	106	130	130	100	120	100	130	100

Level of spontaneous production of IL 1- β: 670 ng ml^{-1}, IL-8: 2169 ng ml^{-1} Level of production induced by LPS of IL 1- β: 250 % of control, IL-8: 310 % of control.

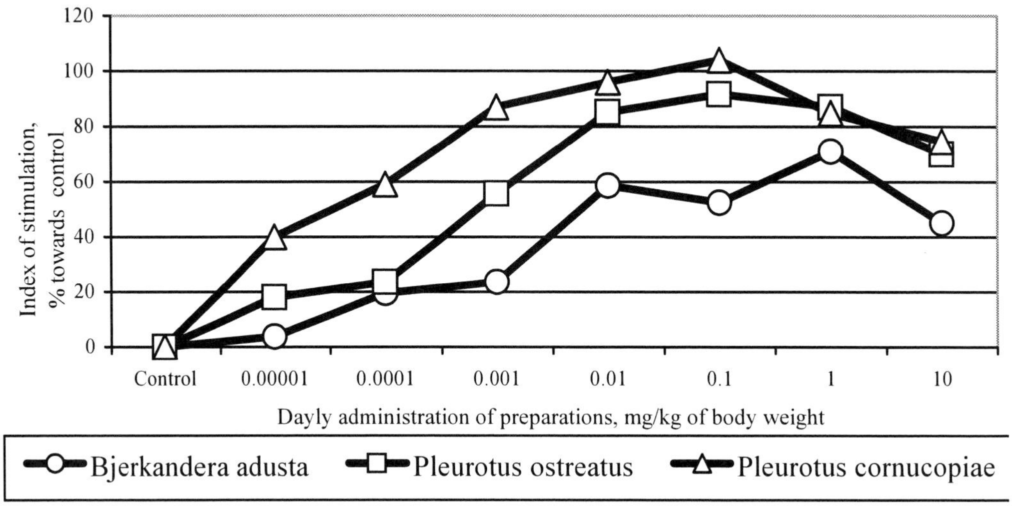

Figure 2. The influence of p.o. administration of mycelia extracts on the levels of antibody forming cells in mice.

The experiments have been done in 16 C57BL female mice (average body weight 23 g) purchased from animal facility "Rappolovo", Russian Academy of Science. Animals were kept 8 per one standard polypropylene chamber. The light regimen was standardized to 12 hours light/dark cycle. All animals received water ad libitum. Animal were divided into two groups – control and experimental. The control group was given usual fodder. The second group animals were given modified fodder with extract (daily dosage 10 mg kg^{-1} body weight in recalculation on a dry extract).

During two-week periods from a beginning of extract administration the average body weight did not differ significantly as compared to control.

At 15 day melanoma B16 was transplanted to all animals (10^6 cells of a melanoma B16 subcutaneously to the right abdomen side). In further measurement of parameters of tumor growth and life duration was conducted.

The progressive and irregular decrease of average body weight was noted in both groups after tumor transplantation but less in group received extract. At 26 day after tumor transplantation average body weight was significantly higher in experimental group. This is the evidence of decrease the body intoxication by mushroom supplements, and thus the increase of the quality of life.

The average tumor size changes are represented in table 3. In control group, average tumor size increased during tumor development. In experimental group average tumor size decrease 30-80 % was marked with maximal effect during first 14 days.

Table 3. The effect of *Pleurotus ostreatus* extract peroral intake on tumour development in mices bearing melanoma B16

Time after tumour transplantation, days	Tumour volume, sm^3		Suppression, %
	Experimental group	Control group	
8	0.02 ± 0.02	0.03 ± 0.01	30
12	0.14 ± 0.09	0.43 ± 0.18	70
15	0.55 ± 0.33	2.19 ± 0.45	75
22	1.52 ± 0.52	4.34 ± 1.24	65
26	1.84 ± 1.30	9.37 ± 2.64	80
29	0.28 ± 0.17	-	-
33	0.11 ± 0.02	-	-
36	0.34 ± 0.01	-	-
40	0.67 ± 0.02	-	-
43	1.22	-	-

The survival rate of mice with a melanoma B16 feed with aqueous extract from pleurotus ostreatus is represented in a figure 3.

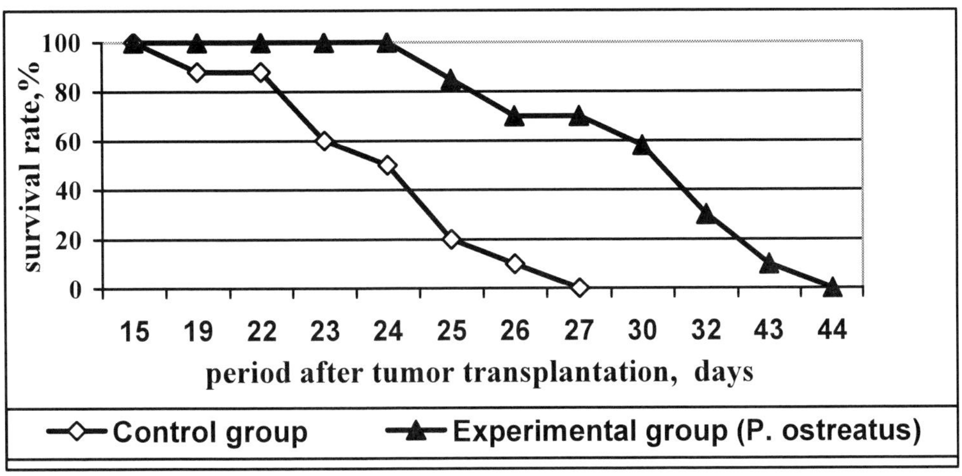

Figure 3. Survival of mice with melanoma B16.

Thus, the extract from *Pleurotus ostreatus* fruit bodies makes reliable inhibiting effect on melanoma B-16 growth and decreases a manifestation of tumor intoxication as it confirms by the weight differences between control and experience mice. These effects perhaps are connected with the stimulation of immune system, which has a large importance for a pathogenesis of the tumor growth.

Experiments on mice with Ehrlich's ascit carcinoma show the oncostatic effect of supplements from another mushroom mycelia – *Cerrena unicolor* (figure 4), while several other species, including *Pleurotus ostreatus* were not efficient in this case.

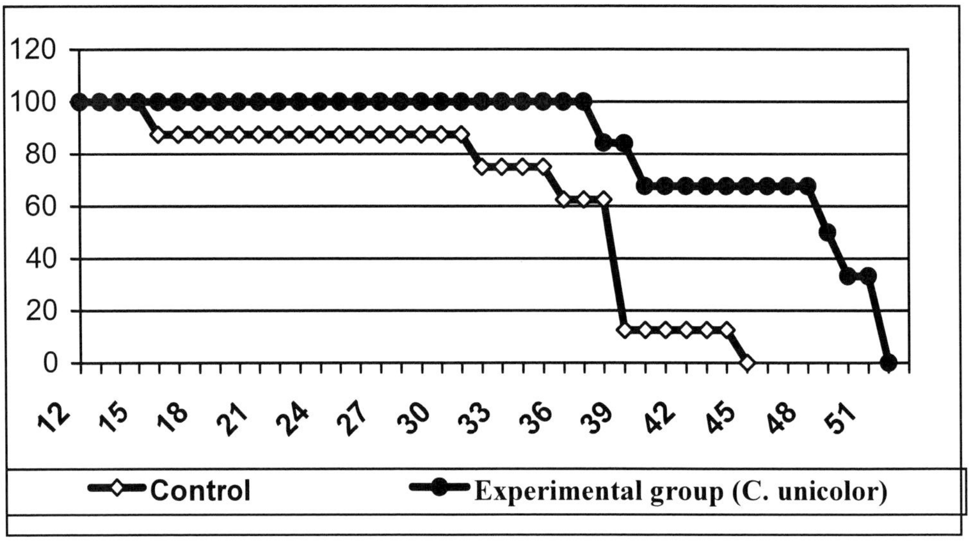

Figure 4. Survival of mice with Ehrlich's carcinoma.

All of the data obtained during our investigations suggest that aqueous extracts of studied Basidiomycetes have marked biological effects. The highest activity was determined for the extracts from *Bjerkandera adusta*, *Pleurotus cornucopiae*, *Pleurotus ostreatus*, *Auricularia auricula-judae*, *Inonutus radiatus*, *Tricholoma portentosum*, *Grifola frondosa* and *Lentinus edodes*. The levels of stimulation of different immunocompetent cells, significantly varies for the extracts of different mushroom species. Probably, this can explain also the different effect towards different tumors, possessing by extracts from various mushroom species.

References

[1] Beiyer, V. A. (1973). Short manual of haematology. *Chimiya,* Leningrad, 118p. (in Russian).

[2] Boyum, A. (1968). Isolation of mononuclear cells and granulocytes from human blood. Isolation of mononuclear cells by centrifugation. *Scand. J. Clin. Labor. Invest.,* 41, 77-89.

[3] Chang, S.T. (1999). Global impact of edible and medicinal mushrooms on human welfare in the 21st century: nongreen revolution. *Int. J. Med. Mushrooms,* 1, 1-7.

[4] Chihara, G. (1992). Immunopharmacology of lentinan, a polysaccharide isolated from Lentinus edodes: its application as a host defense potentiator. *International Journal of Oriental Medicine,* 17, 57-77.

[5] Collins, R.A. and Ng, T.B. (1987). Polysaccharopeptide from Coriolus versicolor has potential for use against human immunodeficiency virus type 1 infections. *Life Science.* 60, 383-387.

[6] Dike, M.M. and Forlin, J.K. (Eds.) *Manual of immunopharmacology. Mir,* Moscow, 1998. 332 pp. (in Russian)

[7] Espenshade, M.A. and Griffith, E.W. (1966). Tumor –inhibiting basidiomycetes. Isolation and cultivation inthe laboratory. *Mycologia,* 58, 511-517.

[8] Gregory, F.J., Healy, E.M., Agersborg, H.P.K. Jr., and Warren, G.H. (1966). Studies of antitumor substances produced by basidiomycetes. *Mycologia,* 58, 80-90.

[9] Hishida, I., Nanba, H. and Kuroda, H. (1988). Antitumour activity exhibited by orally administered extracts from fruit-body of Grifola frondosa (maitake). *Chemical and Pharmaceutical Bulletin.* 36(5), 1819-1827.

[10] Ikekawa, T., (1995). Enokitake, Flammulina velutipes: antitumour activity of extracts and polysaccharides. *Food Review International.* 11, 203-206.

[11] Ikekawa, T., Uehara, N., Maeda, Y., Nakanishi, M., and Fukuoka F. (1969). Antitumor activity of aqueous extracts of edible mushrooms. *Cancer Res*, 29, 734-735 .

[12] Jong, S.C. and Birmingham, J.M. (1992). Medicinal benefits of the mushroom Ganoderma. *Advances in Applied Microbiology.* 37, 101-134.

[13] Kanatsu, N., Terakawa, H., Nakanishi, K., Watanabe, Y. (1963). Flammulin, a basic protein of Flamulina velutipes with antitumor activities. *J. of Antibiotics. Ser. A.,* 16, 139-143.

[14] Liu, G-T. (1999). Recent advances in research of pharmacology and clinical applications of Ganoderma (P. Korst.) species (Aphyllophoromycetideae) in china. *International Journal of Medicinal Mushrooms.* 1, 63-67.

[15] Mizuno, T. (Ed.). (1995). Mushrooms: the versatile fungus-food and medicinal properties. *Food Rev. Int.* (special issue), II, 173-178.

[16] Mizuno, T. (1999). Medicinal effects and utilisation of Cordiceps Fr. Link (Ascomycetes)and Isaria Fd. (Mitosporic Fungi) Chinese caterpillar fungi "Tuchukaso. *Int. J. Med. Mushr,* 1, 251-261.

[17] Mizuno, T. (1999). The extraction and development of antitumor-active polysaccharides from medicinal mushrooms in Japan, *Int. J. Med Mushr,* 1, 9.

[18] Mizuno, T. (2000). Development of an antitumour biological response modifier from Phellinus linteus (Berk. Et Curt.) Teng (Aphyllophoromycetideae) (Review). *International Journal of MedicinalMushrooms.* 2, 21-33.

[19] Nelson, K. D. Chemotaxis under agarose. *J. Immunol,* 115, 1650– 1656 (1975).

[20] Tono-Oro, T., Ueno, N., Matsumoto, T. (1990). Chemiluminescence of Whole Blood, a simple and rapid method for the testimation of phagocytic function of the granulocytes and opsonic activity in whole Blood. *Clin. Immun. Immunopath.,* 26, 66-75.

[21] Wang G., Zhang J., Mizuno T., Zhuang C., Ito H., Mayuzumi H., Okamoto H., Li J., (1993). Antitumor active polysaccharides from Chinese mushroom Songhan lingzhi, the fruiting body of Ganiderma tsugae. *Biocsi. Biotech. Biochem.* 57, 894-900.

[22] Wasser S.P., Weis A. (1999). Medicinal properties of substrates occuring in hygher Basidiomycetes mushrooms: current perspectives. *Int. J. Med. Mushrooms,* 1, 31-62.

[23] Wasser S.P., 2002. Medicinal mushrooms as a source of antitumor and immunomodulating polysaccharides. *Appl. Microbiol. Biotechnol,* Vol. 60, 3, 258- 274.

In: Research Progress in Biotechnology
Editor: G. E. Zaikov, pp. 159-168
ISBN: 978-1-60456-000-8

Chapter XVIII

Mass Spectrometric Approach for Determination of Collagenase Specificity

Yu.V. Sungurov, N.L. Eremeev, A.T. Lebedev, O.A. Maloshitskaya, G.N. Rudenskaya, S.A. Semenova and A.M. Shmoylov
Lomonosov Moscow State University, Chemistry Department.
Leninskie Gory 1, Build. 40, 119992 Moscow, Russia
e-mail: yagry@ramler.ru, tel: +7(495)939-34-17,
fax: +7(495)939-54-17

Abstract

A number of crude collagenase preparations (bacterial metallocollagenase from *Clostridium Histolyticum* and cysteine collagenases *from Serratia Proteomaculans;* cysteine collagenases from larvae *Dermestes frischi and Dermestes Maculatus, and* serine collagenase from the commercial preparation "Morikrase"*)* of varying specificity toward the native rat collagen type I were used to obtain hydrolyzates of collagen which were further analysed by MALDI-TOF MS. Resulted mass spectra may be easily distinguished from each other depending on the enzyme preparation used, while registration of one or another group of peaks in the spectra can be referred only to a certain collagenolytic enzyme preparation. Obtained results permit to propose MALDI-TOF MS analysis of collagen hydrolyzates for the primary determination of collagenase specificity.

Keywords: Collagen/collagenase specificity/peptide mixture/Matrix-assisted laser desoption-ioniztion time of flight mass spectrometry.

1. Introduction

Collagens are a group of related proteins, which have a very high fastness. The fibrillar structure of skin, bones, tendons, cartilages, blood vessels, and teeth are based on them. The peculiarity of amino acid consecution of collagen is well-defined iterative motive Gly-Xaa-Yaa, where Xaa is Pro or HyPro in 75% cases [1].

The main macrostructural unit of collagen fiber is tropocollagen that consists of three helically obvolute strands. Each strand includes 1000 aa. Due to the fact that the tropocollagen chains are mainly bound by hydrogen bonds as well as by covalent crosslinks, collagen exhibits a high resistance to proteolytic enzymes [2].

However, so-called collagenases that can destruct the native collagen under physiological conditions are known among the proteases. There are two extreme types of collagenases in relation of their action on collagen. For example, *Clostridium histolyticum*, a bacterium causing gas gangrene, produces the collagenase that can cleave the polypeptide chain of collagen in over 200 positions (a lot of low molecular weight products are generated) [3-5]. The other type - tissue metallocollagenases that are found in the growing and metamorphosing tissues of amphibians and mammals [6-8], cleave tropocollagen as if cutting all its three chains in the only site: before Gly750 in the sequence of all 1000 aa (two fragments with masses 1/4 and 3/4 of the initial chain mass are resulted). The majority of collagenases has intermediate specificity and form a lot of fragments with different molecular weights in the process of collagen hydrolysis [9, 10]. For nowadays properties of three classes of collagenases: serine, cysteine and metalloproteinase, have been studied. Collagenase containing residues of aspartic or glutamic acids in their active centers are not discovered yet.

The ability of collagenases to destruct collagen fibers is widely used in tanning and fur industry for leather currying [11]; in food industry - for softening of meat and its taste improvement [12]; in fish-processing industry - for dressing of calamaries carcass, for cleaning of caviar from roe [13]; in brewing trade - for the malt quality improving, for clarification and stabilization of beer [13]; in medicine and cosmetology – for treatment of burns, suppurative trophic ulcers, post-surgical scars, osteochondroses etc [14, 15]. In research activities highly purified preaparations of collagenases are applied in carrying out of animal physiology researches, for example, for the particular removing of impeding connective tissue in contact experiments with nerve cells [16].

Collagenases of determined specificity are selected for each specific purpose of application. That is why the screening of new collagenases specificity is very important practical problem. Since the specificity of collagenases determines the assortment of hydrolysis products, the study of collagenase specificity can be realized by the analysis of hydrolyzates composition. In biochemistry such analysis is carried out by the use of isotope labeling and it is quite labor-intensive procedure giving only approximately results [17].

On the other hand methods of mass spectrometric analysis allow us to determine molecular weights of multi-component mixture quickly and easily [18]. One of the modern methods of such analysis that is used as applied to substances of protein nature is a method of Matrix-Assisted Laser Desoption/Ioniztion Time-of-flight Mass Spectrometry (MALDI-TOF MS) [19]. This method makes possible to analyze complex protein-peptides mixtures that are expected after collagenolytic reaction. Since the list of reaction products is quite different for different collagenases, one can propose a method of determination of collagenases action by MALDI-TOF MS analysis of products of their reaction with collagen. For this reason it is necessary to obtain characteristic spectra database for different collagenases and find out the similarity and the difference in these spectra. Such information database (similar to existing databases for tripsinolytic technique in proteomics) can be useful in future for the placing of new collagenases in one or another class.

The aim of presented work is the comparative analysis of mass spectra of hydrolyzates obtained as the result of collagen treatment by collagenases of different types.

2. Materials and Methods

2.1. Materials

Collagen I type isolated from rat tails was kindly submitted by Kalebina T.S. (MSU, Moscow).

Enzyme preparations: clostridiopeptidase A ("Serva", Germany); preparation from *Serratia Proteomaculans*, crude preparations from larvae *Dermestes frischi* and *Dermestes maculatus* were obtained in our laboratory [14]; "Moricrase ("Trinita", Russia).

Acetonitrile and trifluoroacetic acid (TFA) HPLC-grade were purchased from "Kriochrom" (Russia). The MALDI-MS matrices: 3,5-dimethoxy-4-hydroxycinnamic (sinapinic) acid, α-cyano-4-hydroxycinnamic acid were purchased from "Sigma" (USA).

"Peptide Calibraton Standart" and "Protein Calibraton Standart" were purchased from Bruker (Germany).

Salts and components of buffer solutions were purchased from Serva (Germany) and were of the highest available quality.

All solutions were freshly prepared using deionized water purified with a Millipore apparatus (Bedford, MA).

2.2. Carrying out Reactions of Collagen Hydrolysis

Collagen hydrolysis by the different enzyme preparations was carried out at pH 7.5 and t 22 °C in 50 mM Tris-HCl (2.5 mM $CaCl_2$) or 20 mM NH_4HCO_3 buffer solutions. Collagen concentration was 1 mg/ml, enzyme preparation concentration was 0.1 mg/ml. After 24 hours the reaction was stopped by freezing.

2.3. Sample Preparation

For the preparation of analyzed samples for MS detection 1 μl of a reaction mixture was mixed with 1 μl of saturated matrix solution in acetonitrile: water 1:2 vol/vol, containing

0.1% TFA. The resulted mixture was placed on the still target MPT 384 (Bruker Daltonics, Germany) and dried at ambient temperature ("dried droplet" technique).

For sample purification the washing on target was used, which was performed by applying 1-5 μl of washing solution (0.1% aqueous TFA) on the top of preparation, waiting for a few seconds and removing the droplet with pipet.

2.4. MALDI-TOF MS Analysis and Spectra Processing

Samples were analyzed in a time-of-flight mass spectrometer (Autoflex II, Bruker Analytics, Germany). A nitrogen laser (wavelength 337 nm, pulse energy 100 μJ) was used for sample ionization. Positive ion MALDI-TOF mass spectra were recorded in linear or reflection mode and the ion accelerating voltage of 19 kV was applied to the sample holder.

External calibration in different m/z area were performed with the corresponding standards: "Peptide Calibraton Standart" and "Protein Calibraton Standart". Mass spectral analysis for each sample was based on the average of 200-300 laser shots.

Obtained spectra were processed with Bruker DataAnalysis, Version 1.6g ((C) Bruker Daltonik GmbH, 1999). Moniosotopic peaks were manually selected from the set of peaks of molecular ion isotope distribution [20]. Normalized spectra were obtained by the additional processing of spectra - in each spectrum the most intensive peak was set as 100%, and the rest peaks were set in percents as a part of it.

3. Results and Discussion

As it was mentioned above, serine, cysteine and metallocollagenases are known now. Hence, in present work we used enzyme preparations containing collagenases of different classes for obtaining of collagen hydrolyzates for further MALDI-ToF MS analysis. Clostridiopeptidase A from *Clostridium histolyticum* is a metallocollagenase; the main component of enzyme preparations from *Serratia Proteomaculans, Dermestes frischi u Dermestes Maculatus* is different cysteine collagenases; the main component of "Moricrase" is serine collagenase.

Stable mass spectra of substrate, enzymes and reaction mixtures with a good resolution of peaks corresponding to low molecular weight products of hydrolysis were obtained using α-ciano-4-hydroxicinnamylic acid as a matrix (figure 1). Reliable peaks with m/z > 3200 in mass spectra both native collagen and its hydrolyzates were not obtained in experiment by using of different matrix. Possibly this fact can be concerned with properties of collagen structure. As we said earlier collagen fibrils consist of three super stranded chains and stabilized by a lot of intermolecular hydrogen and covalent bonds. It can lead to forming the high-dense polymer net during the process of cooperative crystallization (drying) of the probe. Such process can block the desorption of high-molecular weight products from the target.

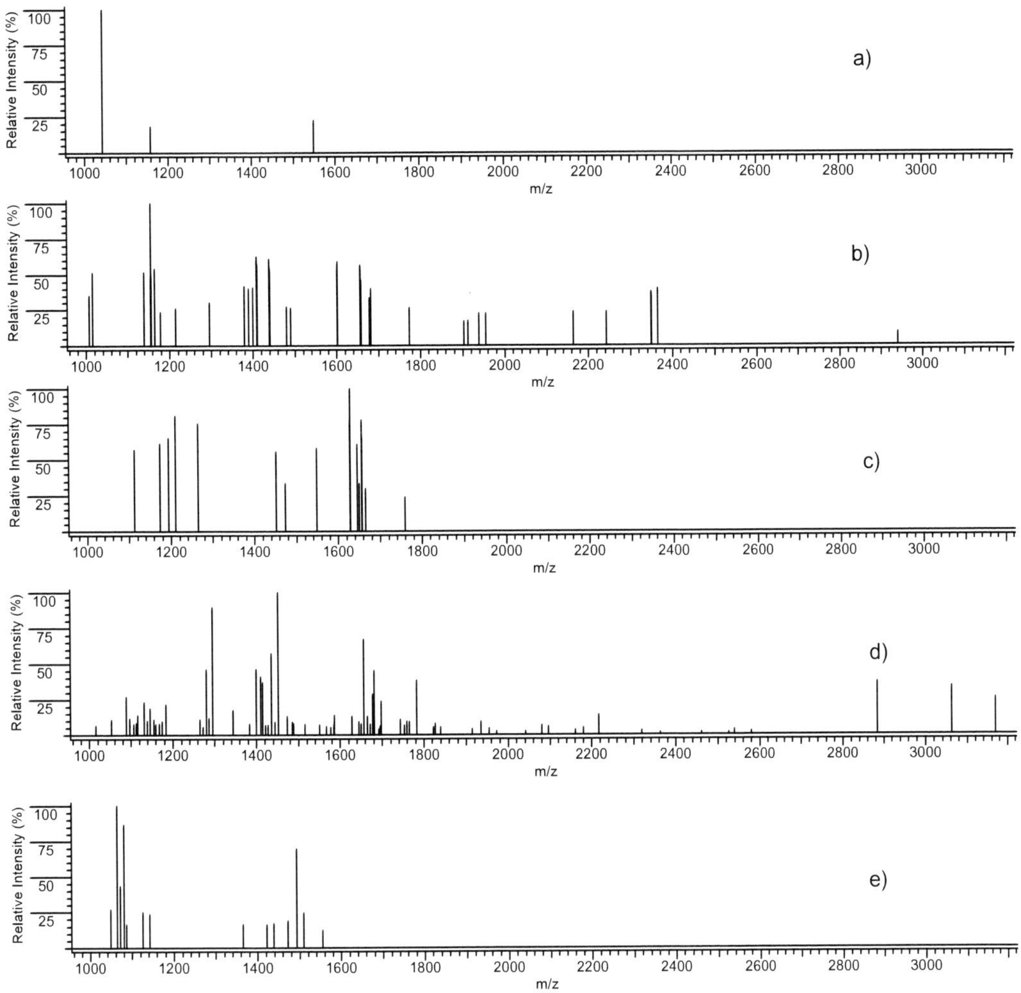

Figure 1. Mass spectra of collagen hydrolyzates (1000<m/z<3200) obtained under the action of Clostridiopeptidase (a), preparation from *Serratia Proteomaculans* (b), crude preparations from *Dermestes frischi* (c) and *Dermestes maculatus* (d), "Moricrase" (e).

The m/z value equal to 1000 was chosen as a low bound of m/z range for the mass spectra analysis. The lower mass region of the mass spectrum is congested, filled with ions arising from the matrix molecules, and from degradation of the matrix molecules on interaction with the laser beam, and obviously not takes into account during the mass spectra of complex peptide mixtures processing [21],

Most of collagenases require the presence of calcium ions in a reaction mixture for effective hydrolysis of collagen. On the other hand it is well known that the presence of salts and buffer components can form adducts with analyzed molecules and pollute obtained spectra [18]. In order to clarify the potential influence of reaction mixture components on obtained mass spectrum quality we used the washing procedure of dried on the target probes. Washing of samples was used according to recommendation of mass spectrometer manufacturers. However, there were no changes in mass spectra in concerned range for samples before and after washing.

Another experiment related to potential influence of calcium and TRIS ions on obtained mass spectrum quality was carried out with "Moricrase". Cysteine collagenase from this preparation doesn't require the presence of calcium ions for its collagenolytic activity. So, we carried out the collagen hydrolysis by "Moricrase" in the absence of calcium in ammonium-hydrocarbonated buffer used in proteomics for protein primary structure detection after trypsin digestion [22]. Mass spectrum of hydrolyzate in such buffer solution was identical for mass spectrum of hydrolyzate obtained from TRIS-HCl buffer solution containing calcium ions. These experiments permit to say that in obtained spectra there are no peaks of buffer solutions and/or adducts of this components with peptide ions. Therefore, it means that detected peaks shown in figure 1 are real proved peaks corresponding to low molecular weight products of collagen hydrolysis.

In spite of evident differences of obtained mass spectra in each compared pair of spectra few common peaks always exist. The degree of spectra similarity, i.e. occurrence of similar peaks in 2 mass spectra, is different for each pair. Thus, for mass spectra of collagen hydrolyzates obtained under the action of collagenolytic enzymes of different classes the mass spectra identity achieved 35%, but for collagen hydrolyzates obtained under the action of collagenolytic enzymes of the same class (cysteine collagenases) it achieved 77%.

Such similarity of obtained mass spectra can be defined by the presence in substrate (collagen) iterative motive Gly-Xaa-Yaa with the high percentage of Pro or HyPro in Xaa position. This means that during the collagen hydrolysis a lot of peptides with the different amino acid sequences but the same m/z value (not distinguishable by simple MALDI technique) can appear. For instance, only $\alpha 1$ collagen chain contains 16 variants of possible peptides with m/z 1015 (this m/z presents in many spectra). Of course, identification of the products can be done by MS-MS procedure. But such technique is incompletely suitable for proline-contained peptides due to the high complexity for the interpretation of proine fragmentation spectrum [23].

Nevertheless if we take off all peaks that are similar even to one pair of compared mass spectra all of them demonstrate the presence of peaks specific only for concrete sample (table). The presence of such characteristic peaks in mass spectra (i.e. hydrolytic products with the distinct masses) can be referred only to the action of one or another collagenolytic preparation.

Mainly MALDI technique is not used in quantitative aspects because of many standardization procedures necessity [24, 25]. But it is evidently that the intensity of molecular ion peak depends on the concentration of such spice in analyzing probe. Of course, at the same time the peak intensity depends on the other, non-extensive factors [18]. Nevertheless at the same conditions one can assume that the peaks with the greatest intensity in mass spectrum are referred to the molecular ions with the greatest concentration in the probe.

Based on such assumptions we compared mass spectra of collagen hydrolyzates consisted of the peaks with relative intensity more than 50% (figure 2). The results are very interesting: mass spectra of probes obtained with enzymes of different classes (serine, cysteine and metallocollagenase) do not contain the similar peaks. It means that the set of the hydrolytic products with the high concentration in the probe is unique for the enzymes of different classes. Such differences allow to conclude that peaks with the high intensity can be considered as the peculiar collagenase "finger-print" and can be used for the characterization of enzyme action on collagen.

Table. M/z values of characteristic peaks present in mass spectra of collagen hydrolyzates obtained under the action of different enzyme preparations

Preparation	m/z
Clostridium histolyticum	1042, 1158
Serratia Proteomaculans	1007, 1066, 1123, 1164, 1178, 1214, 1379, 1390, 1481, 1773, 1938, 2164, 2242, 2348
Dermestes frischi	1194, 1210, 1704, 1707
Dermestes Maculatus	1053, 1097, 1116, 1132, 1146, 1160, 1183,1272, 1280, 1286, 1344, 1383, 1415, 1428, 1436, 1444, 1485, 1515, 1567, 1577, 1584, 1586, 1671, 1695, 1698, 1752, 1765, 1823, 1826, 1839, 1935, 1972, 2041, 2080, 2096, 2160, 2180, 2216, 2318, 2461, 2526, 2539, 2580, 2884, 3062, 3169
Paralithodes Camtschatica	1048, 1064, 1080, 1126, 1142, 1365, 1493, 1509, 1555

However, another fact is in the interest too – mass spectra of probes obtained after the treatment of collagen with the enzymes from the same class demonstrate the presence of similar peaks. All mass spectra of collagen hydrolyzates obtained under the action of cysteine collagenases contain the peak with m/z 1656, not presented in mass spectra of hydrolyzates obtained with the enzymes of other classes. One more similar peak (m/z 1452) appears in mass spectra of hydrolyzates, obtained with the using enzymes from the similar sources – cysteine collagenases from larvae Dermestes. Evidently such similarities are determined by the accumulation of the same reaction products under the collagen treatment with enzymes with the closely related specificity.

The obtained results, of course, have now only preliminary character. It is necessary to expand the number of collagenases for obtaining mass spectra of collagen hydrolyzates. The presence of required set of mass spectra allows to create database containing the characteristic peaks referred to different enzyme preparation.

The searching both the individual peaks and their combination in such a database can easily give the answer about the similarity of one or another mass spectrum with the mass spectra contained in the database. The analysis of similarities and differences can answer the question about the character of enzyme action on collagen.

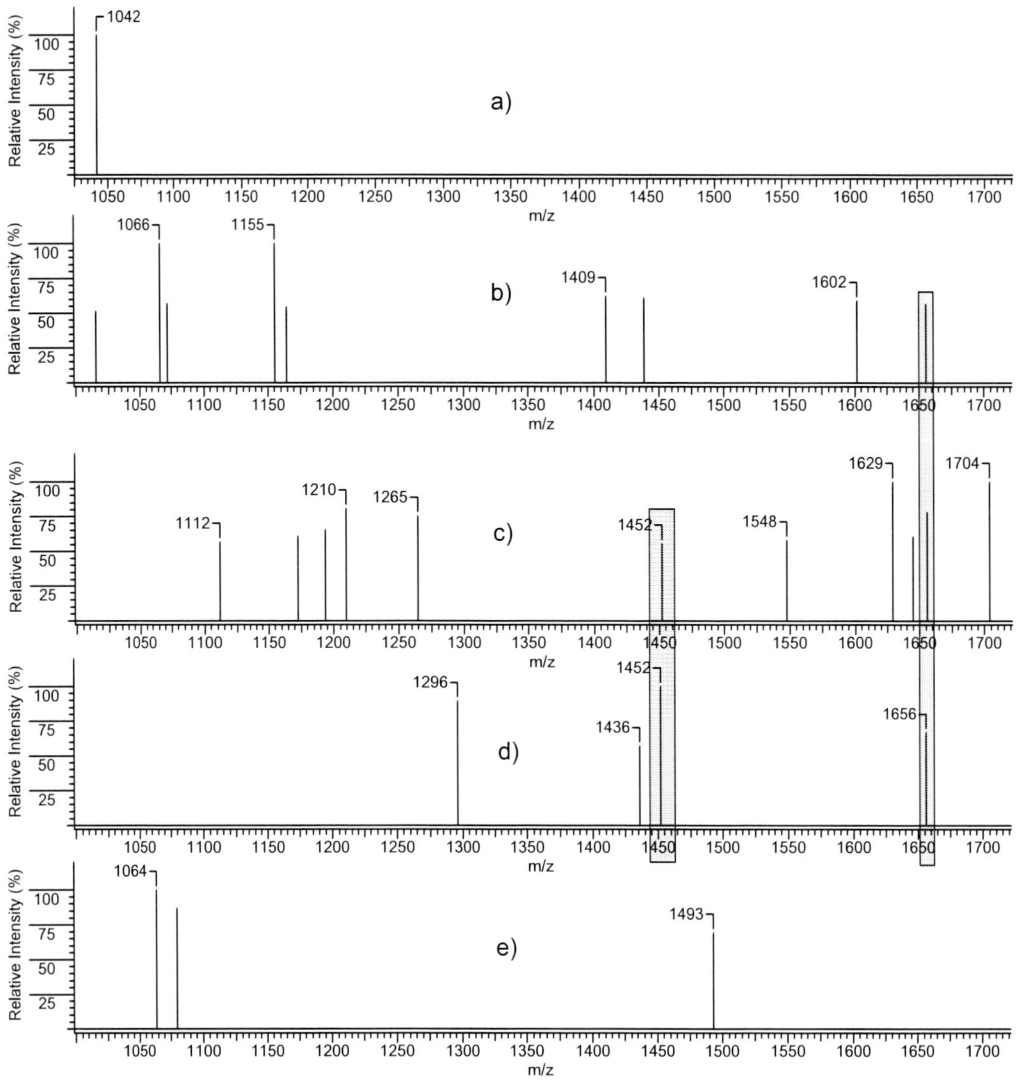

Figure 2. Mass spectra of collagen hydrolyzates (obtained under the action of Clostridiopeptidase (a), preparation from *Serratia Proteomaculans* (b), crude preparations from *Dermestes frischi* (c) and *Dermestes maculatus* (d), "Moricrase" (e)) wich contain peaks with relative intensives more than 50%.

4. Conclusion

Obtained results show that MALDI-TOF MS technique can be used for the investigation of collagenolytic reaction products. Moreover, our study shows the potential applicability of such approach for the primary determination of collagenase specificity in reaction with collagen.

References

[1] V. Nikitin., E. Perskiy, L. Utevskaya: *Naukova Dumka,* Kiev, 1977. 269 p. (In Russian).
[2] V.M. Stepanov (Ed.): Molecular Biology (structure and functions of proteins). *High schools,* Moscow, 2002. 288 p. (In Russian).
[3] M.D. Bond, H. Van Wart: *Biochemistry.,* 23, 3092 (1984).
[4] S.K. Mallya, K.A. Mookhtiar, H. Van Wart: *J. Protein. Chem.,* 11, 99 (1992).
[5] I. Mandl, J. D. MacLennan, E. L. Howes, R. H. DeBellis, A. Sohler: *Clin. Invest. J.,* 32(12), 1323 (1953).
[6] H. Birkedal-Hansen, W. Moore, M. Bodden, L. Windsor, B. Birke-dal-Hansen, A. DeCarlo: *Crit. Rev. Oral Biol. Med.,* 4, 1197 (1993).
[7] H. Kolkenbrock, T. Wieder, N. Ulbrich: *Biol. Chem. Hoppe Seyler.,* 372(2), 83 (1991).
[8] B. Lovejoy, A. Hassell, M. Luther, D. Weigl, S. Jordan: *Biochemistry,* 33(7), 8207 (1994).
[9] G.A. Grant, J.C. Sacchettini, H.G. Welgus: *Biochemistry,* 22, 2228 (1983).
[10] O.A. Klimova, S.I. Borukhov, N.I. Solovyeva, T.O. Balaevskaya, A.Y. Strongin: *Biochem Biophys. Res Commun., 166,* 1411 (1990).
[11] L.V. Antipova, Ch.Yu. Shamkhanov, C.E. Mishin, A. R. Goz: *Modern Advances of BoiTechnology,* Stavropol, 2002, 100 p. (In Russian).
[12] C. McDonagh, E. M. Desmond, F. J., Monahan, D. J. Troy: *Irish Journal of Agricultural and Food Research,* 39(1), 138 (2000).
[13] G. A. Macdonald, K. A. Hofman: *Native collagen from fish skin,* European Patent EP1233985, 2002.
[14] G. N. Rudenskaya, A. M. Shmoilov, A. V. Baydakov, S. A. Semenova, D. D. Kharlampieva, V. A. Isaev: 3rd Moscow International Congress *«Biotechnology: state of the art and prospects of development»,* Moscow, 2005, 18 p.
[15] A. M. Shmoilov, G. N. Rudenskaya, , A. V. Baydakov, R. D. Zhantiev, O. S. Korsunovskaya, L. V. Ageeva, N. V. Starikova: *J. Drug. Del. Sci. Tech.,* 16(4), 285 (2006).
[16] O. A. Klimova, S I. Borukhov, N. I. Solovyeva, T. O. Balaevskaya, A Ya.Strongin: *Biochem. Biophys. Res. Commun.,* 166(3), 1411 (1990).
[17] A. E. Berman, V.V. Bychkova, G.E. Morozevitch, V.I. Mazurov: *Voprosy med. Khimii.,* 30(1), 108 (1984) (In Russian).
[18] A. T. Lebedev: Mass Spectrometry in Organic Chemistry. *Binom,* Moscow, 2003, 493 p. (in Russian).
[19] K. Tanaka, H. Waki, Y. Ido, S. Akita, Y. Yoshida, T. Yoshida: *Rapid Commun. Mass Spectrom.,* 2, 151 (1988).
[20] J. Yergey, D. Heller, G. Hansen, R. J. Cotter, C. Fenselau: *Anal. Chem.,* 55, 353 (1983).
[21] Busch K.L.: *J. Chromatogr.* A,692, 275 (1995)
[22] R. Aebersold, D Goodlett: *Chem. Rev.,* 101(2), 269 (2001).

[23] M.F. Ocaca, J. Jarvis, R. Parker, P. M. Bramley,J. M. Halket, R. K. Patel, H. Neubert: *Proteomics,* 5(5), 1209 (2005).
[24] D. Immler, S. Greven, P. Reinemer: *Proteomics,* 6(10), 2947 (2006).
[25] M. Quadroni, A. Ducret, R. Stocklin: *Proteomics,* 4(8), 2211 (2004).

In: Research Progress in Biotechnology
Editor: G. E. Zaikov, pp. 169-172
ISBN: 978-1-60456-000-8

Chapter XIX

Two Enzyme Immunoassay for Diphtheria Toxin Detection

Sh.S. Tashmukhamedova*
National University of Uzbekistan named M. Ulugbek
Campus, Tashkent 700174, Uzbekistan

Abstract

The system for determination of diphteria toxin on the base of liposomal firm - phase immunoassay analysis was developed. For the detection of the signal is concerned two-enzyme system. The effect of the conjugate with phospholipase was enhanced with amylase encapsulated in multylammellar liposomes. The optimal parameters immunoassay analysis have been estimated and scheme of running of the liposomal immunoanalysis is presented.

Keywords: Immuno - enzyme assay, antigens, antibody, liposome, homogeneous and heterogeneous methods, complement.

Immuno - enzyme assay is one of informative methods of analysis of natural chemical compounds that widely used in research practice as well as in medical diagnostics of different antigens [1]. At the same time, permanently is conducted searching for the new methods for the expansion of immuno - enzyme assay means (resources, opportunities). One of the such a new approaches was become the liposomal immunoassay [2-3]. Using of the liposomes was made possible the development of the new homogeneous [4-5] and heterogeneous methods of analysis [6]. In this case, such methods more often based on the lytic ability for complement which activated after formation of antigen - antibody [7-8] complex or on the inhibition activity of the antibodies that became apparent after reaction with the Mellitin conjugates [9].

* Sh.S. Tashmukhamedova: E-mail: t_meroj@mail.ru, fax: +998371 1447917

In so doing, the enzyme marker was encapsulated in liposome [10]. In the presented work described the system of liposomal firm - phase analysis on polyamide micropore membrane (film). For the registration of signals during immuno - enzyme assay was used the two - enzyme system that enhances the sensitivity of analysis. This was account for by the action one enzyme - phospholipase A_2 which part of conjugate was enhance by the second enzyme – α-amylase encapsulated in liposomes.

Immune reaction was carried out in two stages. First one is blocking the part of antibodies by certain antigens molecules. Plentiful antibodies coupled with the enzyme were added to the pre washed antigen - antibody complex. Conjugates of antibodies with enzyme are able to make contacts only at the point where the molecules of antigens are present. This signifies that the amount of antigens in the probe determine the amount of antibody - enzyme complex and determine the level of enzyme activity at the stage of substrate - enzyme reaction. The next stage was washing of the carrier, adding the substrate and detection the amount of the enzyme coupled with the immunosorbent. The antigen between immobilized antibody and antibody coupled with the enzyme named "sandwich"-complex.

The amount of liposome contained the indicator - enzyme was the significant value for the liposomal immunoassay. In this connection, the immunoassay experiments with different liposome concentrations and stable amount of conjugates were conducted. Optimal liposome dilution was 1:2 times (figure 1,a), but for the next steps of optimization was used 1:5 dilutions, in that case the amount of marker enzyme was sufficient for signal detection.

The increasing of substrate concentration above optimal caused decreasing of enzymatic activity because of substrate inhibition. This is evident by the data of enzyme immunoassay conducted with different substrate concentrations and most hydrolyzed was 0.5% substrate (figure 1, b).

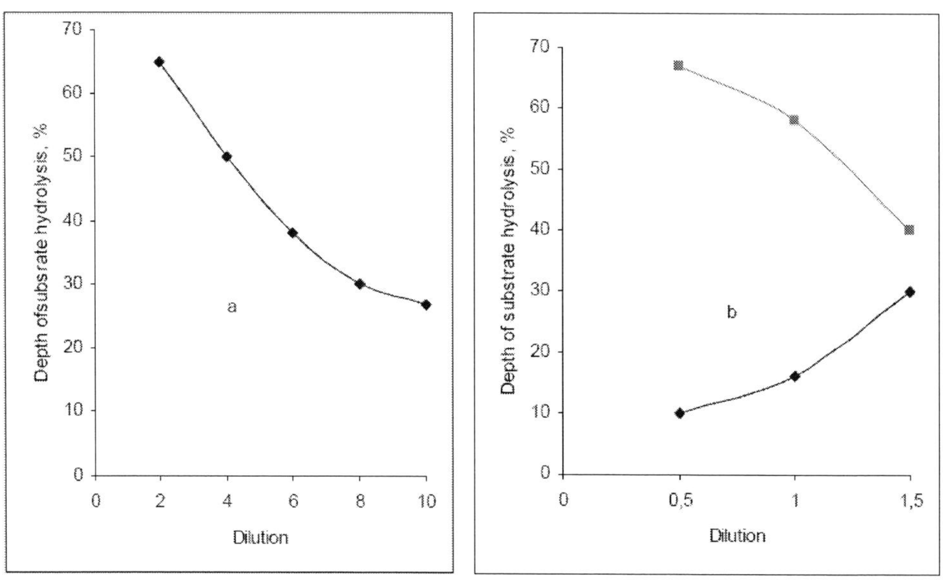

Figure 1. Dependence of depth of substrate hydrolysis on different parameters used in LILA. a – influence of liposome concentration (liposome dilution); b – influence of substrate concentration.

For finding out what time essential for specific enzyme-substrate reaction, the time dependence of enzymatic activity of phospholipase A_2 and б-amylase was studied. After 5 min of reaction whole amylase released from liposome and therefore, there was no use in further increasing the time of interaction of phospholipase A_2 with liposome. The data of study time dependence of б-amylase reaction showed that enzymatic activity increased proportionally to the reaction time, but amount of depth of starch hydrolyzes was enough in 10 min of reaction.

Development of the test-system for the detection diphtheria toxin aims. Test-system was a two - enzyme version of "sandwich"- assay. The role the substrate for phospholipase A_2, which was a part of conjugate were played the liposomes with encapsulated б-amylase. After lysis of liposome by phospholipase A_2 б-amylase released out. In next stage, the substrate for б-amylase – starch was added and by detecting depth of starch hydrolyzes the amount of marker coupled with immunosorbent was determined.

The procedure of assay was simple, there was no need in the special equipment and consisted the systematic immersing membrane with immobilized antibodies to the several solutions. Before starting the assay, the membrane with immobilized antibodies was washed up and was equilibrated with 0.05M Tris - HCl buffer, pH 6.0, containing 30 mM $CaCl_2$ and 5% Ethylenglycol. Then 5ng/ml diphtheria anatoxin was added. Incubation time was 30 min under continuous mixture. Within this time, the antigens were bound specifically with immobilized antibodies. After incubation membrane nee to be washed up from uncoupled antigens and 1 ml 0.05 M Tris-Hl buffer, pH6.0, containing 35 mkg of conjugate antibody with phospholypase-A_2 was added and incubated during 30 min. During the incubation, antibodies of conjugates make a complex with antigens on the membrane.

The next stage was adding the liposome and mixture them during 5 min, 37°C. Within this time, phospholipase-A_2 of conjugate completely broke up the membrane of liposome and caused б-amylase releasing. Then 0.2 ml of 0,5% starch added and after 10 min of hydrolysis at 37°C, 0.1 ml product of hydrolysis transferred to the 10 ml of Iodide solution. Depth of starch hydrolysis determined at the Photoelectrocalorimeter.

The amount of antigen in sample was estimated through the calibration curve. For development of calibration curve, purified diphtheria anatoxin in concentration from 0.005 to 5000 ng/ml was used. Enzymatic activity of conjugate was proportional to the concentration of antigen, which was evidence of keeping enzymatic and antigen activity of conjugate. The lowest amount of antigen detected by this method was 0.05 ng/ml.

Thus, there were determined optimal conditions of testing the products of enzymatic reaction on the liposomal enzyme immunoassay on the membranes with covalently immobilized antibodies using phospholipase-A_2 as marker enzyme and б-amylase as indicator enzyme.

References

[1] D.S. Hage : *Anal.Chem.* 1999.71.p.294

[2] H.A. Rongen, A. Bult, W.P. Bennekom: *J. Immunol. Methods.,* 1997. 204. p.105

[3] J.H. Howanitz : *Arch Pathol Lab Med.* 1988. 204. p.775

[4] K. Hosoda, T. Yasuda: *J. Immunol Methods.* 1989. 121. p.121

[5] E.F. Ullman, T. Tarnowsky, P. Felgner, I. Gibbons: *Clin.Chem.* 987. 33. p.1579.

[6] S. Tretyakov, S. Yarkov, V. Zlobin, Y. Kalinin: *J. Microbiol. epidemiol. and immunobiology.* 1992. 11-12. p.43

[7] M. Umeda, T. Yasuda: *Acta Med. Okayama.* 1994. 48. p.299

[8] G. Akots, J. Braman, R. Broeze, D. Bowden: *Complement.* 1984. 1. p.125

[9] J. Freytag, W. Litchfield: *J. Immunol. Methods.,* 1984. 70. p.133

[10] K. Kubotsu, S. Goto, M. Fujita, H. Tuchiya, M. Kida, S. Takano, S. Matsuura, I. Sakurabayashi: *Clin Chem.* 1992. 38. p.808

In: Research Progress in Biotechnology
Editor: G. E. Zaikov, pp. 173-181
ISBN: 978-1-60456-000-8

Chapter XX

Anaerobic Solid-State Fermentation of Vegetative Substrata by Probiotic Strain *Bacillus Subtilis*

N.A. Ushakova**, *E.S. Brodskij, B.A. Chernuha, A.A. Kozlova, A.V. Nifatov and D.S.Pavlov
Severtsov Institute of Ecology and Evolution Russian Academy of Sciences 119071. Moscow

Abstract

Anaerobic solid-state fermentation of vegetative substrata by probiotic strain *Bacillus subtilis* 8130 is used for reception of highly effective biologically active additives to forages of agricultural animals. A basis of biotechnology - biochemical transformations of vegetative polymers under influence of bacilla, similar to the processes occurring in rumen of ruminant animals. On the data of chromate-mass-spectrometric analysis among products of fermentation organic acids (acetate, propionate, and butyrate), their derivative and low-molecular aromatic connections are found out.

Keywords: anaerobic solid-state fermentation, cellulose, probiotic, Bacillus subtilis.

One of actively developing directions of microbe's biotechnology is development of probiotic preparations for animals - the alive microbe's food additives influencing an organism by improvement of its microbes balance, stimulating exchange and immune processes [1-7]. The urgency of a problem of reception effective probiotic is determined by requirements of modern industrial animal industries, poultry farming for growth factors of animals, and connected to deterioration of ecological, sanitary-and-epidemiologic conditions

* N.A. Ushakova: Moscow, the Lenin prospect, 33; ushakova@sevin.ru; Fax (495) 954-55-34

in the world on a background of the tendency of a society to restriction of application in forages of antibiotics.

Fundamental bases of creation of such preparations are developed insufficiently. One of debatable questions is the scientific substantiation of application of bacteria of sort *Bacillus* for reception industrial probiotic preparations. And if legitimacy of use *Lacto-* and *Bifidobacterium* in quality probiotic can be explained by their resident character, i.e. constant presence at digestive system of animals [8,9] that concerning representatives *Bacillus* a question it is more difficult. If they also are found out in an intestinal tract, their amount is a small [10-12]. However many modern probiotic preparations effectively used in the treatment-and-prophylactic purposes, contain bacteria of one or two strain *Bacillus* [13]. Most frequently probiotic preparations include strain *B.subtilis.* It is possible to explain the given fact rich metabolic opportunities and adaptable plasticity of representatives of this kind of bacteria in a combination to features of their ecology.

The wide circulation of bacteria *Bacillus subtilis* in the nature (on plants and in carrion vegetative rests, in ground, water) leads to natural contact to an animal organism. With forage they can get in a digestive tract. Important property *B.subtilis* is participation in initial stages of cellulose degradation [14]. *Bacillus* is found out in departments of digestive path herbivorous vertebrate animals in whom there are basic processes anaerobic digestion cellulose [15]. As humidity of contents, for example, a blind gut, does not exceed 75 % [16], and the oxidation-reduction potential makes -200 мВ [17], the intestinal fermentation of components of a forage, mainly, vegetative polymers, is carried out solid-state fermentation in anaerobic conditions. Cells of *B.subtilis* can develop in such conditions, and carry out a mix type of carbohydrate's fermentation [18].

The purpose of the present work was modeling anaerobic solid-state fermentation of vegetative substrata with the raised contents of cellulose (vegetative waste products) with the help natural intestinal strain *B.subtilis* 8130, and studying of formed products of fermentation.

Materials and Methods

In work used natural intestinal strain *B.subtilis* 8130 with probiotic properties [19]. Aerobic deep cultivation of bacilli carried out in nutrient media, involved corn meal, barmy extract, mineral salts; pH 7.0. As vegetative substrata for solid-state process have been taken sea-buckthorn husk, sunflower husk and beet bin cake. Solid-state fermentation carried out at humidity of weight of 70 % in the sealed - in polyethylene packages in conditions of the complicated access of oxygen.

For studying products of anaerobic solid-state fermentations with the help chromate-mass-spectrometric analysis received dichloromethane extracts of products of a fermentation of the specified types of vegetative substrata. As the control similar extracts of a mix of a sterile liquid nutrient medium with corresponding sterile vegetative raw material without entering bacilli served. In addition investigated extracts of a liquid nutrient medium and nutrient liquid after development of *B.subtilis* 8130. Analyses made on chromate-mass-spectrometry system including gas chromatograph Trace GC Ultra and the mass -

spectrometer detector « ionic trap » POLARIS Q (THERMOFINNIGAN). Conditions of the analysis: a quartz capillary column 30м x 0.25 mm with motionless phase DB-5ms (layer 0.25 microns), programming of temperature from 50^0C up to 220^0C with a speed in 10^0C/minutes, then up to 280^0C with a speed in 4^0C/minutes. Temperature injector 260^0C, the interface - 250^0C, ionization by electronic impact at energy of electron - 70 eV, scanning of mass - spectra from 41 up to 550 a.e.м. with a speed 1 spectrum in 1 second. Entered 1 mkl a solution in a mode of division of a stream 1:10. Identification of components carried out a method of an expert estimation on spectra-structural correlations with use of library search.

The product of anaerobic solid-state fermentations of sea-buckthorn husk manufactures of Joint-Stock Company "Altaivitamine" was mixed with fodder of broilers and with the received experimental forage fed skilled chickens in Open Company "Novoelovskaia" an integrated poultry farm, Altai territory. Control broilers fed with similar mixed fodder with addition of an antibiotic Flavomicin ("Intervet").

Results and Their Discussion

The complex structure of the substrata, which was exposed anaerobic solid-state fermentation, was reflected in the common structure of mass-chromatograms. At comparison of experience with the control disappearance on mass-chromatograms skilled preparations of some peaks present in the control, and occurrence new was found out. On a difference of components in spectra judged occurrence of the products connected to metabolic activity of bacilli. Mass-chromatograms of preparations and corresponding controls with the indication of the identified components of a bacterial origin are submitted on figures 1-3.

Strain *B.subtilis* 8130 it is capable at anaerobic solid-state fermentation of all investigated types of vegetative substrata to form acetate, propionate, butyrate, various derivatives of butyrate, such, as 2-methyl-, 3-methyl butyrate, nonsaturated derivatives. Besides on spectra aromatic connections and derivatives of furan are identified. Such substances can have bactericidal an effect, and play a part in biological efficiency of the preparations received at anaerobic solid-state fermentation. Products of fermentation *B.subtilis* 8130 are submitted in table 1. It is necessary to note, that the basic metabolites, formed at cultivation of bacilli in aerobic conditions in a liquid nutrient medium, are submitted ethyl acetate, diethyl acetate that does not render essential influence on structure of products of anaerobic solid-state fermentation of vegetative substrata. Biochemical transformations of vegetative polymers under influence of *B.subtilis* in many respects are similar to the processes occurring in ruminant [20]. The structure of products of a bacterial metabolism is influenced type of the vegetative substratum, exposed with fermentations, pH the reactionary environment, time of carrying out of process. However formation of the basic products of a fermentation - low-molecular organic acids - under identical other conditions of conducting process essentially does not depend on biochemical features of cellulose a substratum.

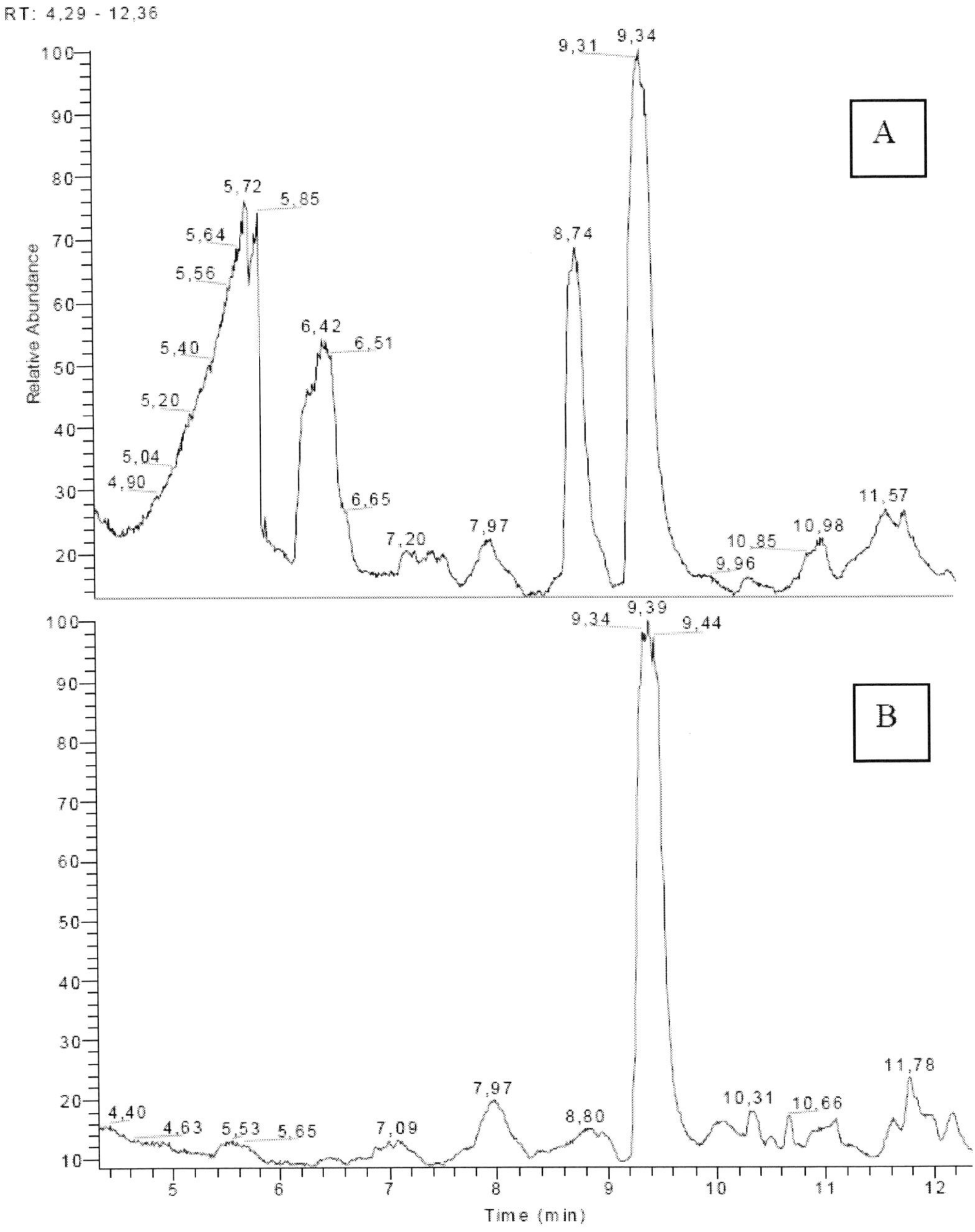

Figure 1. An initial part of mass- chromatogram of products of a sea-buckthorn husk anaerobic solid-state fermentation by *Bacillus subtilis* 8130 (A) and the control (B). The identified components of a bacterial origin: RT:5.72 - acetate, RT:5.85 - propionate, RT:6.42 - butyrate, RT:8.74 – an C_6 acid, RT:11.57- benzoic acid.

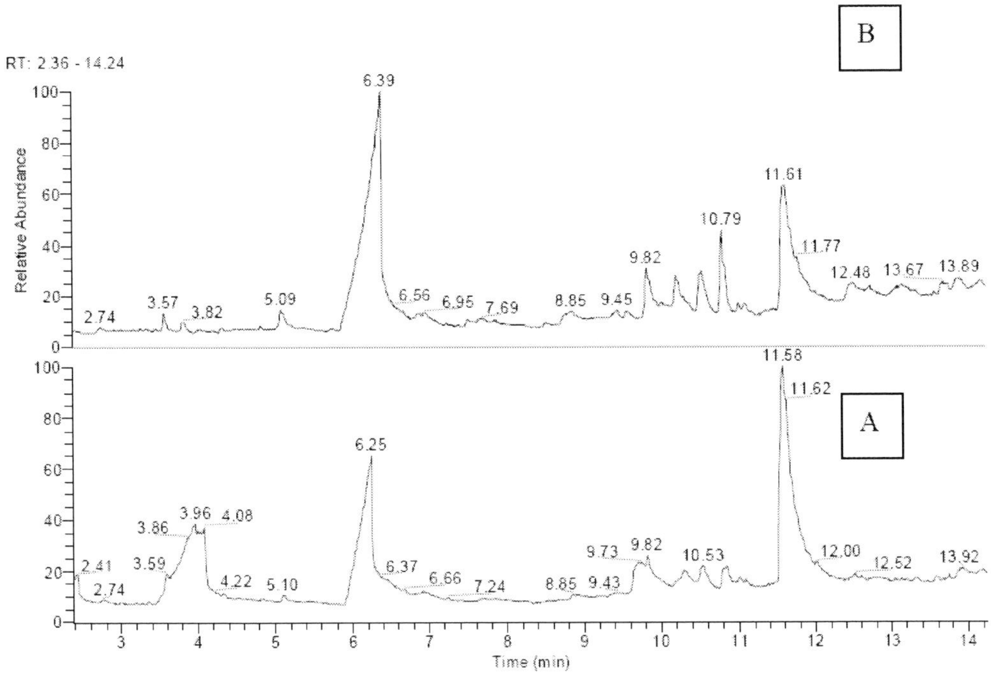

Figure 2. A comparative kind of initial sites of mass- chromatogram of products of anaerobic solid-state fermentation of a beet bin cake (A) and the control (B). The identified components of a bacterial origin: RT: 3.96-4.22 - 2-methyl-, 3-methyl butyrate and not identified ether; RT:9.73 - 2-furancarboxaldehyde, 5-(hydroxymethyl).

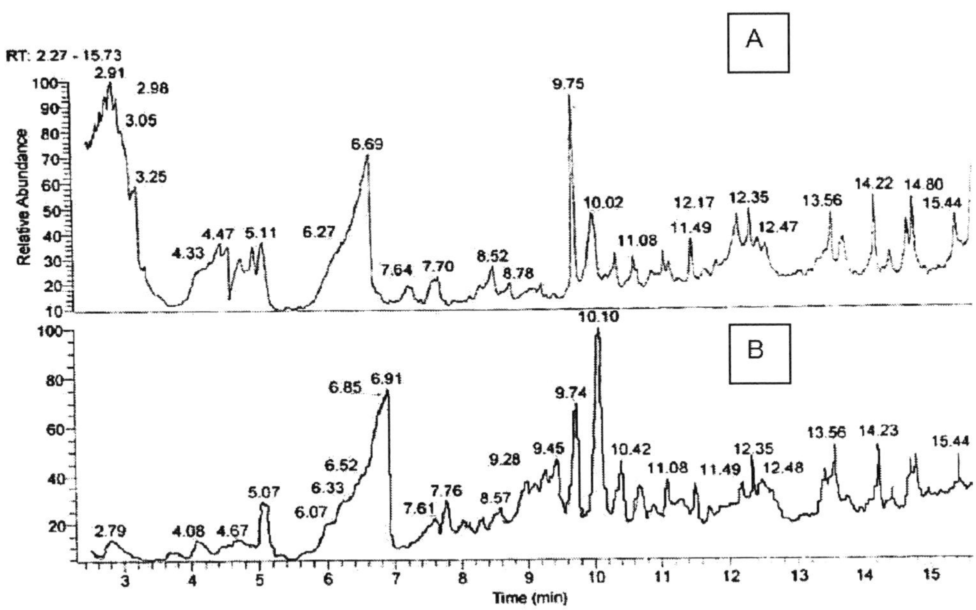

Figure 3. A general view of an initial part of mass- chromatogram of products of a sunflower husk anaerobic solid-state fermentation (A) and the control (B). The basic identified components of a bacterial origin: RT:4.47 - 3-methyl butyrate; RT:4.57 - 2- methyl butyrate; RT:5.11 - furan, 2-pentyl.

The anaerobic solid-state fermentation of vegetative substrata by probiotic strain *Bacillus subtilis* 8130 has been applied for reception of biologically active additives to forage of agricultural animals. Results of experience on broilers, into which diet entered a product anaerobic solid-state fermentations of sea-buckthorn husk (skilled group), in comparison by control chickens (control group) to which gave a fodder antibiotic flavomicin, are submitted on table 2. The daily average additional weight of skilled broilers was higher control 4.4 % that has allowed receiving in addition more than 100 g from a head. Safety of a livestock has been comparable to the control receiving an antibiotic. Experiment has shown high biological efficiency of a product of anaerobic solid-state fermentation of a vegetative substratum that, probably, is connected to biotechnological modeling processes of digestion cellulose of a forage and reception biologically active secondary metabolites that is typical for solid-state fermentation [21].

Table 1. The identified products of anaerobic solid-state fermentation of a sea-buckthorn husk, sunflower husk, a beet bin cake by *B.subtilis* 8130

Substance	Structure
2- Methyl butyrate	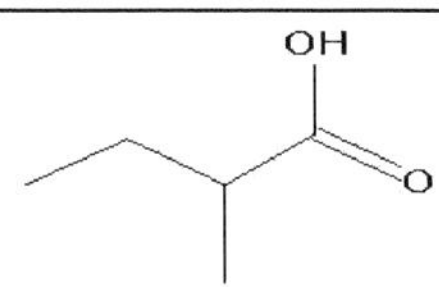
3-Methyl butyrate	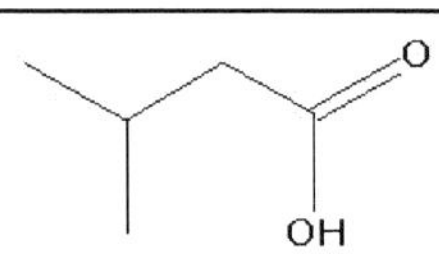
3-Butenoic acid, 3-methyl-	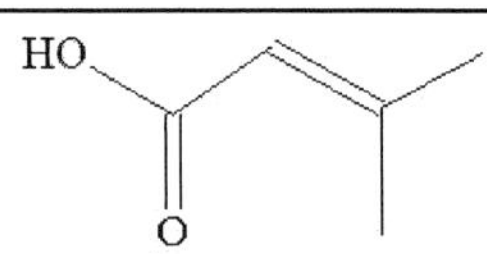
Oxalic acid	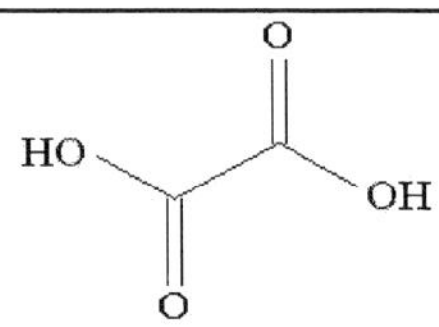
Valeric acid, 4-pentadecyl ester	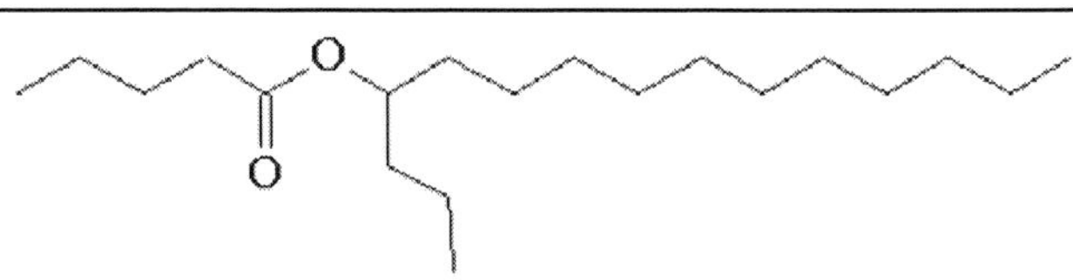
Benzoic acid	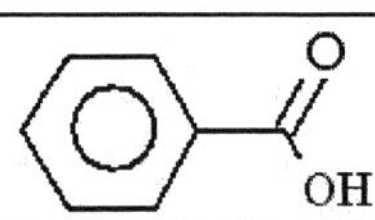

Table 1. (Continued).

Substance	Structure
Acetobenzoic acid	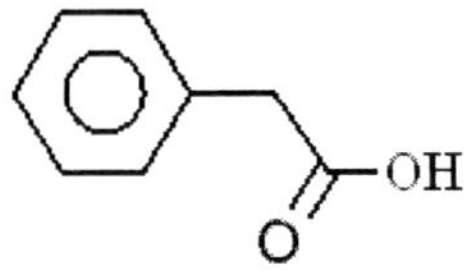
Phenyl - 2- hydroxypropionate	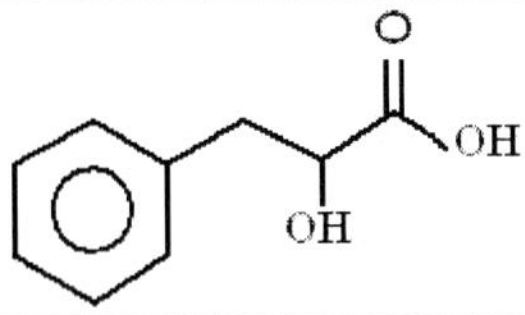
Phenylpropyl alcohol	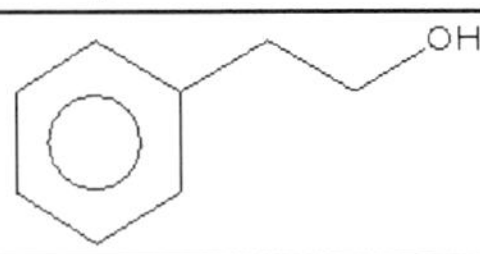
Phenethyl alcohol	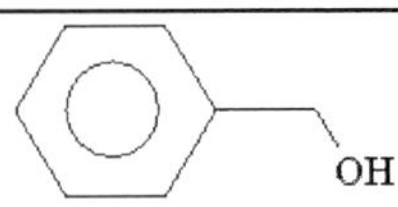
2-Furancarboxaldehyde, 5-(hydroxymethyl)-	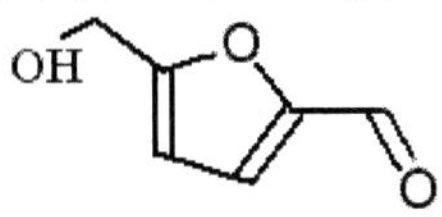
Ethylamine	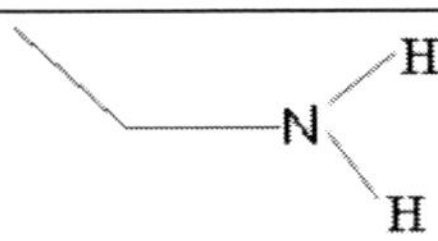
2- Hydroxyethylamine	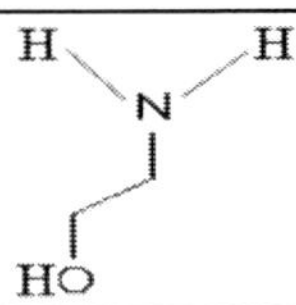
3-Buten-2-one, 4-[4-(dimethylamino)phenyl]-	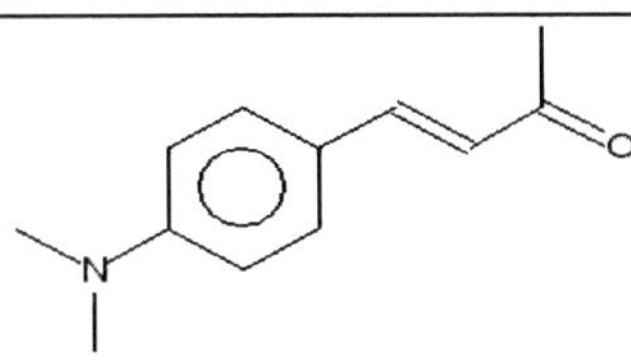
Furan, 2-pentyl-	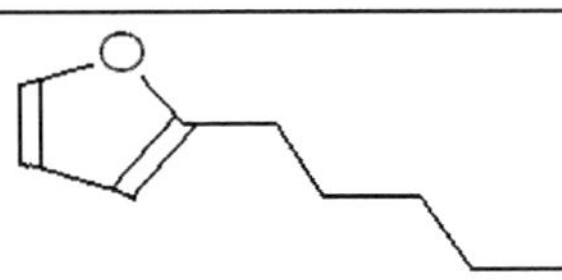

Table 1. (Continued).

Substance	Structure
2(3H)-Furanone, 5-heptyldihydro-	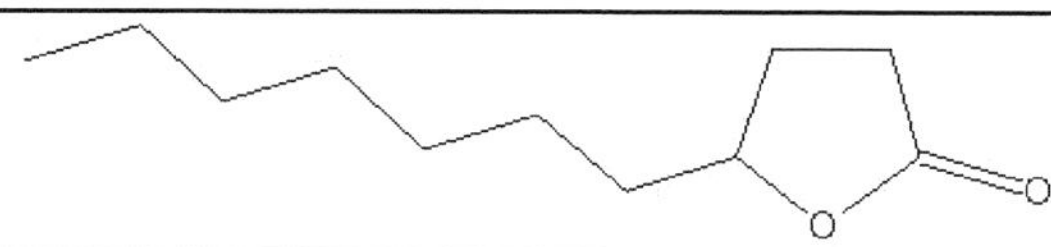
2-Oxoglutaric acid	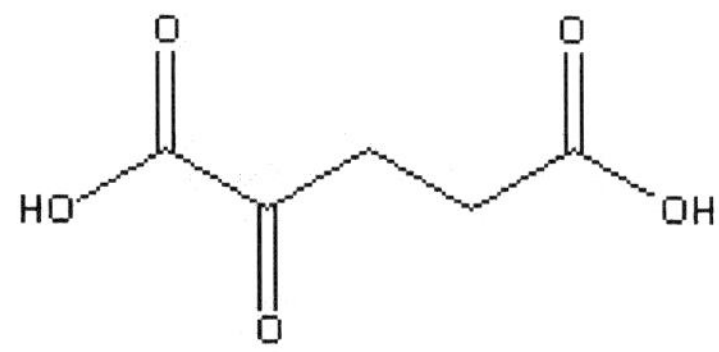

Table 2. Parameters of efficiency fattening broilers, Cross IZA-15

Parameter	Groups	
	Control	Skilled
Number of individuals	24820	28943
Age at slaughter, days	42	42
Safety, %	93.4	93.0
Daily average additional weight, g	45.4	47.4
Average weight of 1 head, g	1941	2054

References

[1] R.Fuller (Ed.): *Probiotics. The scientific basis.* Chapman and Hall. London. N.Y. Tokyo. 1992, 397p.

[2] M.D. Collins: *Am. J. Clln. Nutr.* 1999, 69, 1052S.

[3] B.V.Cockroach: *Veterinary science.* 2000, 47.

[4] M.B. Roberfroid: *Am. J. Clin Nutr,* 2000, 71(6), Suppl. P.1682-1687.

[5] N.I.Malik, A.N.Panin: *Veterinary science.* 2001,1, 46, (in Russian).

[6] B.Zimmermann, E.Bauer, R.Mozenthin: *J. Anim. Feed. Sc.* 2001, 10(1), 47.

[7] D. Bonn: The Lancet Infectious Diseases, 2002, 2, 716.

[8] G. Reid: *Appl. Environ. Microbiol.* 1999, 65(9), 3763.

[9] N.J.Kashirskaja: *the Russian medical magazine,* 2000, 8 (13-14), 572, (in Russian).

[10] H.Bergner, H.A.Kejtts: *Scientific bases of a feed of agricultural animals. Mir,* Moscow, 1973, 597, (in Russian).

[11] I.G.Pivnjak, B.V.Cockroach: *Microbiology of digestion ruminant. Mir,* Moscow, 1982, 248, (in Russian).

[12] A.A.Vorobjev, E.A.Lykov: *Microbiological magazine,* 1999, 6, 102, (in Russian).

[13] L.F.Bakulina, I.V.Timofeev, N.G.Perminova, etc.: *Biotechnology,* 2001, 2, 48, (in Russian).

[14] Bergey's *Manual of Systematic Bacteriology:* Williams and Wilkins. 1986, 2, 1104.

[15] N.A.Ushakova, L.P.Belov, A.A.Varshavsky, A.A.Kozlova, T.V.Kolganova, E.S.Bulygina, T.P.Turova: *Microbiology,* 2003, 72(3), 400, (in Russian).
[16] J.A.Kalugin: Physiology of a feed of rabbits. Kolos, Moscow, 1980,175, (in Russian).
[17] R.A.Celesk, T.Asano, M.Wagner: *Pros. Soc. Exp. Biol. and Med.* 1976, 151, 260.
[18] J.W.Lengeler, G.Drews, H.G.Schlegel (eds): *Biology of the Procaryotes. Mir,* Moscow, 2005, 1, 380, (in Russian).
[19] N.A.Ushakova, A.A.Kozlova, A.V. Nifatov: *Biotechnology,* Agriculture and the Food Industry. Ed. G.E.Zaikov. Nova Science Publishers, Inc. New York, 2006, 111.
[20] P.J. Van Soest: *Nutritional Ecology of the Ruminant.* (2nd end). Ithaca, NY:Cornell Univ. Press, 1994, 367 p.
[21] T.Robinson, D.Singh, P.Nigam: *Appl. Microbiol. Biotechnol.* 2001, 55, 284.

Index

A

D

G

H

I

K

L

M

N

O

P

Q

R

S

T

U

V

W

Y